Community Forestry in Nepal

Community forestry focuses on the link between forest resources and livelihoods and contributes to forest conservation and reforestation. It is widespread in Nepal, with a very high proportion of the rural population involved, and is widely recognized as one of the most successful examples of community forestry in Asia. Through a combination of literature reviews and original research, this volume explores key experiences of community forestry in Nepal over the last four decades as a model for improving forest management and supporting local livelihoods.

The book takes a critical approach, recognizing successes, especially in forest conservation and restoration, along with mixed outcomes in terms of poverty reduction and benefits to forest users. It recognizes the way that community forestry has continued to evolve to meet new challenges, including the global challenges of climate change, environmental degradation and conservation, as well as national demographic and social changes due to large-scale labour migration and the growing remittance economy. In addition to examining the changes and responses, the book explores ways that community forestry in Nepal might move forward. Lessons from Nepal have relevance to community forestry and community-based approaches to natural resource management around the world that are also experiencing global pressures and opportunities.

Richard Thwaites is Senior Lecturer, School of Environmental Sciences, and Institute for Land Water and Society, Charles Sturt University, Australia.

Robert Fisher is Senior Lecturer, School of Geosciences, University of Sydney, and Senior Research Fellow, Tropical Forests and People Research Centre, University of the Sunshine Coast, Australia.

Mohan Poudel is REDD Expert (Under Secretary), REDD Implementation Centre, Ministry of Forest and Soil Conservation, Nepal.

The Earthscan Forest Library

This series brings together a wide collection of volumes addressing diverse aspects of forests and forestry and draws on a range of disciplinary perspectives. Titles cover the full range of forest science and include the biology, ecology, biodiversity, restoration, management (including silviculture and timber production), geography and environment (including climate change), socio-economics, anthropology, policy, law and governance. The series aims to demonstrate the important role of forests in nature, peoples' livelihoods and in contributing to broader sustainable development goals. It is aimed at undergraduate and postgraduate students, researchers, professionals, policy-makers and concerned members of civil society.

Series Editorial Advisers:

John L. Innes, Professor and Dean, Faculty of Forestry, University of British Columbia, Canada.
John Parrotta, Research Program Leader for International Science Issues, US Forest Service - Research & Development, Arlington, Virginia, USA.
Jeffrey Sayer, Professor and Director, Development Practice Programme, School of Earth and Environmental Sciences, James Cook University, Australia, and Member, Independent Science and Partnership Council, CGIAR (Consultative Group on International Agricultural Research).

Community Forestry in Nepal
Adapting to a Changing World
Edited by Richard Thwaites, Robert Fisher and Mohan Poudel

Forest Governance and Management Across Time
Developing a New Forest Social Contract
Erland Mårald, Camilla Sandström and Annika Nordin and others

Sustainable Forest Management
From Concept to Practice
Edited by John L. Innes and Anna V. Tikina

Additional information on these and further titles can be found at http://www.routledge.com/books/series/ECTEFL

Community Forestry in Nepal

Adapting to a Changing World

Edited by
**Richard Thwaites, Robert Fisher
and Mohan Poudel**

Routledge
Taylor & Francis Group
LONDON AND NEW YORK

from Routledge

First published 2018
by Routledge

2 Park Square, Milton Park, Abingdon, Oxfordshire OX14 4RN

52 Vanderbilt Avenue, New York, NY 10017

Routledge is an imprint of the Taylor & Francis Group, an informa business

First issued in paperback 2019

British Library Cataloguing-in-Publication Data
A catalogue record for this book is available from the British Library

Library of Congress Cataloging-in-Publication Data
Names: Thwaites, Richard (Earth sciences professor), editor. | Fisher,
   Robert (Earth sciences professor), editor. | Poudel, Mohan, editor.
Title: Community forestry in Nepal : adapting to a changing world /
   edited by Richard Thwaites, Robert Fisher, and Mohan Poudel.
Description: Abingdon, Oxon ; New York, NY : Routledge, 2018. |
   Series: The Earthscan forest series | Includes bibliographical references
   and index.
Identifiers: LCCN 2017034106| ISBN 9781138214620 (hardback) |
   ISBN 9781315445168 (ebook)
Subjects: LCSH: Community forestry—Nepal. | Community
   forestry—Himalaya Mountains Region. | Forests and forestry—Nepal.
   | Forests and forestry—Himalaya Mountains Region. | Forests and
   forestry—Social aspects—Nepal. | Forests and forestry—Social
   aspects—Himalaya Mountains Region.
Classification: LCC SD654.6.N47 C66 2018 | DDC
   333.75095496—dc23
LC record available at https://lccn.loc.gov/2017034106

ISBN: 978-1-138-21462-0 (hbk)
ISBN: 978-0-367-40372-0 (pbk)

Typeset in Goudy
by Apex CoVantage, LLC

Contents

Figures

Tables

Boxes

Figure 0.1 Mosaic landscape in Lalitpur District in the central Middle Hills zone, showing dryland paddy and forest mix

(Source: R. Thwaites)

Figure 0.2 Production of dried fruit leather (candy) from *lapsi* fruit (*Choerospondias axillaris*) owned by Community Forest User Group in Sindhu Palchok District (Chapter 5)

(Source: B. Devkota)

Foreword

It gives me an immense pleasure to have been asked to provide a Foreword by the editors of this book, *Community Forestry in Nepal: Adapting to a Changing World*, which includes scholars who have contributed significantly to the understanding of forestry from socio-cultural as well as biophysical perspectives. One of its editors/contributors, I must reveal, is fondly remembered by students, teachers and researchers in Nepal as one of the authors of the seminal book on community forestry titled *Villagers, Forests and Foresters* (Gilmour and Fisher 1991). I say this in order to let the reader know that I do not need to introduce the editors or the authors of the papers included in the present book (as is often expected in a Foreword). This volume is likely to remain as popular because of the wide range of issues and problems the authors have covered in examining the role of community forestry in Nepal's attempts at reconciling conservation and development. The range of topics and issues covered in the book is an acknowledgment of the fact that there has been a continuous evolution in community forestry's agenda (see also Kanel 2004) and that newer issues and challenges are likely to come up in the future.

The authors in the book are suggesting (on the basis of empirical data) that community forestry has been resilient to the changing political, economic and environmental issues of today's socio-culturally and economically intertwined and globalizing world. This is the story of dealings that has come to our notice ever since the community forestry program was initiated in the late 1970s. The situation in the past was different and we now know well that narratives on Nepal's forests, environment and communities have taken various forms at different time periods even during the recent past. For instance, not long ago, the farmers in the hill villages were implicated for an impending environmental degradation in the Himalayan region (see Ives and Messerli 1989 for details and a rejoinder to this discourse; see also Griffin 1988). The same hill farmers who were seen by some as causing problems to environmental health were discovered by others to be taking initiatives on their own in protecting and managing forests closer to their respective villages (see Fisher 1989; Chhetri and Pandey 1992; Chhetri 1993). In fact some communities reported that they had taken such initiatives from around the beginning of the twentieth century (see Chhetri and Pandey 1992; Chhetri 1993). If we look closely at today's Community Forest User Group model (close to 20,000 CFUGs have been formally recognized by the state), it becomes evident that community-based natural resources including forest management initiatives of the past had inspired the re-discovery of such models (see Fisher, 1989; Gilmour, 1990). That is, the discovery of indigenous management systems was of critical importance for the Community Forest User Group model to take off.

Our understanding of the reality on the ground has constantly improved ever since the community forestry program was initiated in Nepal while research-based information has enabled us to eulogize the very village communities (hill farmers) as custodians of natural resources who, until not long ago, were seen as the guilty party for causing deforestation. Had we looked at how forests were treated as sources of revenue by the state while forests were owned by the state, community and private individuals, we would have been more cautious before pointing our fingers at anyone. An economic historian, M.C. Regmi (on the basis of legal documents and available reports) tells that during the Rana rule in the country, the state itself was engaged in export of timber and other forest products (see Regmi 1988). Thus we see that there was a time period when the state authorities regarded forests as sources of timber and non-timber products which could fetch income to the state treasury.

Today, when local people have been entrusted with the protection and management of forests in the country, there is not just an improvement in the condition of the forest, but also the total area of the forests in the country is now reported to have increased. What factors must be responsible for this to happen? The chapters in the book do provide some answers, while answers to similar questions were given in another important volume that came out as the proceedings of a workshop titled '25 Years of Community Forestry'. There can be many more reasons that we have not learned about yet and future research and deliberations may find their answers.

This book, I would say, is unique in that it opens a space for discussion of climate change, REDD+, etc. in a single book that also includes discussions on issues of tenure, livelihood outcomes, community development, etc. Community forestry need no more be limited to talking about issues such as conservation, local development, gender and inclusive decision-making, etc. only. This justifies the subtitle of the book because it tells clearly that community forestry is adapting to a changing world and not just to its custodians, managers or users in the villages. Moreover, it is an apt characterization of how community forestry has changed over the years – while CF itself remains a catalyst for transformations in the social, economic and human aspects from the hill villages to the national level.

In my reading, what the authors of the chapters in this book are trying to do is to contend that natural resource management with a variety of models around the world is an evolving practice that has kept pace with developments on issues, questions and debates at the global level – in relation to how resources are understood, dealt with and managed. To my knowledge this is the first book on community forestry in Nepal that includes papers on climate change and REDD+ together with discussions on biophysical and socio-economic aspects of forest management. This only suggests how important climate change and REDD+ are becoming to the governments and donor agencies in today's globalized community.

This book is likely to stimulate further research on other forms of community-based natural resource management from socio-economic as well as cultural and political perspectives. The important thing for me is that this book may encourage future researchers to look at local resource management institutions from many angles and by using different methods, whether their interests happen to be biophysical, conservation, commercial or economic, cultural, religious or political.

Ram B. Chhetri
Professor of Anthropology,
Tribhuvan University, Kathmandu
September 29, 2017
Kathmandu, Nepal

References

Chhetri, R.B. 1993, Indigenous protection and management systems of forests in the Far Western Region of Nepal, in D. Tamang, G.J. Gill and G.B. Thapa (eds), *Indigenous Management of Natural Resources in Nepal*, Ministry of Agriculture/Winrock International, Kathmandu, Nepal, pp. 323–342.

Chhetri, R.B. and Pandey, T.R. 1992, *User Group Forestry in the Far-Western Region of Nepal (Case Studies from Baitadi and Achham)*, International Centre for Integrated Mountain Development, Kathmandu.

Fisher, R.J. 1989, *Indigenous Institutions and Organizations in the Management of Common Property Forest Resources in Nepal*, East-West Center, Environment and Policy Institute Working Paper No. 18, Honolulu, Hawaii.

Gilmour, D.A. 1990, Resource availability and indigenous forest management systems in Nepal, *Society and Natural Resources*, vol. 3, pp. 145–158.

Gilmour, D.A. and Fisher, R.J. 1991, *Villagers, Forests and Foresters: The Philosophy, Process and Practice of Community Forestry in Nepal*, Sahayogi Press, Kathmandu.

Griffin, D.M. 1988, *Innocents Abroad in the Forests of Nepal: An Account of Australian Aid to Nepalese Forestry*, Anutech, Canberra.

Ives, J.D. and Messerli, B. 1989, *The Himalayan Dilemma: Reconciling Development and Conservation*, Routledge, London and New York.

Kanel, K.R. 2004. Twenty five years of community forestry: Contributions to millenium development goals, in K.R. Kanel, P. Mathema, B.R. Kanel, D.R Nirula, A.R. Sharma and M. Gautam (eds), *25 Years of Community Forestry: Contributing to Millenium Development Goals*, Proceedings of the Fourth National Workshop on Community Forestry, pp. 4–18. Community Forestry Division, Department of Forests, Kathmandu.

Regmi, M.C. 1988, *An economic history of Nepal—1846–1901*, Nath Publishing House, Varanasi.

Preface

The idea for this book arose from the research of four students from Nepal, each undertaking their PhD research at Charles Sturt University in Australia. All four grew up in rural villages in Nepal and were graduates of the Institute of Forestry at Tribhuvan University, and chose to investigate topics related to the experience of community forestry in Nepal. With considerable experience in the Nepal Department of Forests, and the implementation of community forestry, Binod Devkota began his research in Australia in 2008 with the objective of investigating the socio-economic outcomes of community forestry in Nepal. Mohan Poudel and Popular Gentle both began their studies in 2011. With many years of experience working in the NGO sector on community development, Popular Gentle was interested in the contribution of local institutions, particularly community forestry, to the capacities of local rural communities to adapt to climate change. After some years in the Department of Forests of Nepal as a district forest officer and as a planner for the implementation of REDD+, Mohan Poudel came to Australia to investigate the outcomes of the implementation of REDD+ through community forestry, and the implications for local people. Eak Rana came to Australia in 2012 from the International Centre for Integrated Mountain Development (ICIMOD) in Nepal, where he was the project coordinator of the REDD+ pilot project in Nepal. His research was focused on the dynamics of interactions between local livelihoods, forest biodiversity and carbon sequestration, to explore trade-offs and synergies from the implementation of the REDD+ program through local community forestry institutions.

The combination of their PhD research and their extensive experience on the ground in Nepal, involved in implementation of the community forestry program, and its links with forest management, local livelihoods and poverty alleviation and institutional governance offered a rich opportunity for exploration through a book. We have brought together these highly experienced practitioners from the field of forestry, community forestry and community development, with a small number of other experts in related fields, to tell the story of community forestry in Nepal, its past history, successes and challenges, but also the changing context of community forestry and challenges that arise from that. This is told largely through Nepali eyes, and with a strong research evidence base. We believe that this story has relevance beyond Nepal, to practitioners and students of community forestry approaches throughout the world, as well as to other participatory and community-based approaches to natural resource management.

The original PhD dissertations are referenced in the chapters as Devkota (2012), Gentle (2014), Poudel (2014) and Rana (2016).

These research dissertations can be found through the following weblinks:

Devkota, Binod Prasad. 2012, *Socio-economic outcomes of community forestry in Nepal: Lessons from three diverse rural communities*, PhD, viewed 16 May 2017: http://primo.unilinc.edu.au/primo_library/libweb/action/dlDisplay.do?vid=CSU2&docId=dtl_csu40107

Gentle, Popular. 2014, *Equipping poor people for climate change: Local institutions and pro-poor adaptation for rural communities in Nepal*, PhD, viewed 16 May 2017: http://primo.unilinc.edu.au/primo_library/libweb/action/dlDisplay.do?vid=CSU2&docId=dtl_csu59617

Poudel, Mohan Prasad. 2014, *Examining outcomes of REDD+ through community forestry in rural Nepal*, PhD, viewed 16 May 2017: http://primo.unilinc.edu.au/primo_library/libweb/action/dlDisplay.do?vid=CSU2&docId=dtl_csu75928

Rana, Eak. 2016, *REDD+ and ecosystem services trade-offs and synergies in community forests of central Himalaya, Nepal*, PhD, viewed 16 May 2017: http://primo.unilinc.edu.au/primo_library/libweb/action/dlDisplay.do?vid=CSU2&docId=dtl_csu88032

Acknowledgements

The editors would like to acknowledge the efforts of the contributors to this book for their original research and ongoing commitment to understanding a diversity of complex issues relevant to community forestry. Without the enthusiasm of these contributors, we could not have drawn together their experience and their complementary but distinct stories into this volume. We also gratefully acknowledge the essential support of our respective academic institutions and employers – Charles Sturt University, the University of Sydney and University of Sunshine Coast – and the REDD Implementation Centre in the Ministry of Agriculture and Soil Conservation, Nepal. We are further indebted to our colleagues within these institutions for their intellectual stimulation and support. Rik Thwaites would specifically like to thank the Institute for Land, Water and Society at Charles Sturt University for their financial support of various aspects of this research and its publication, through scholarship and project funds.

For production of the book, we are indebted to the staff at Routledge, particularly to Tim Hardwick and Amy Johnston for their patience and encouragement, and to Autumn Spalding of Apex CoVantage for technical assistance. We also acknowledge the willing assistance of Deanna Duffy of the Charles Sturt University Spatial Data Analysis Network (SPAN) for production of maps and plates.

Finally we are deeply indebted to our families for their ongoing encouragement and support. Mohan Poudel wishes to thank Neelam Karki (Poudel) for her strong support and continued encouragement. Bob Fisher thanks Wilma Fisher for her support and her long-term commitment to Nepal. Rik Thwaites thanks Wendy Connor for her understanding, inspiration and unfailing encouragement.

Acronyms and abbreviations

ANSAB	Asian Network for Sustainable Agriculture and Bio-resources
CBFM	Community-based forest management
CDM	Clean Development Mechanism
CF	Community forestry
CFUG	Community Forest User Group
CFUG EC	Community Forest User Group Executive Committee
CIFOR	Centre for International Forestry Research
COP	Conference of Parties (of UNFCCC)
CSO	Civil society organization
DFID	Department for International Development
DFO	District Forest Office
DFRS	Department of Forest Research and Survey
DoF	Department of Forests
ERP	Emission Reduction Program
FAO	Food and Agriculture Organization
FCPF	Forest Carbon Partnership Facility (of World Bank)
FECOFUN	Federation of Community Forestry Users, Nepal
FGD	Focus group discussion
FOP	Forest operational plan
GoN	Government of Nepal
ha	Hectare
HJCFUG	Helejaljale 'Ka' Community Forest User Group
ICIMOD	International Centre for Integrated Mountain Development
ICS	Improved cooking stove
IGA	Income generating activity
IPs	Indigenous peoples
kg	Kilogram
KCFUG	Kankali Community Forest User Group
KP	Kyoto Protocol
LAPA	Local Adaptation Plan of Action
MEDEP	Micro-Enterprise Development Program
MFSC	Ministry of Forests and Soil Conservation
MRV	Monitoring, Review and Verification system

NPK	Nitrogen, Phosphorus and Potassium
NAPA	National Adaptation Program of Action
NGO	Non-government organization
NORAD	Norwegian Agency for Development Cooperation
NRM	Natural resource management
NRs	Nepalese rupees
PES	Payment for environmental services
REDD	Reducing Emissions from Deforestation and Forest Degradation
REDD+	Reducing Emissions from Deforestation and Forest Degradation and enhancement of carbon stocks
REDD IC	REDD Implementation Centre
R-PIN	REDD Readiness Project Idea Note
RRI	Rights and Resources Initiative
RWG	REDD Working Group
SDCFUG	Shreechhap Deurali Community Forest User Group
UNDP	United Nations Development Program
UNFCCC	United Nations Framework Convention on Climate Change
USD	United States Dollars
VAT	Value added tax
VDC	Village Development Committee

Contributors

Dr Krishna Adhikari is a Research Fellow at the Institute of Social and Cultural Anthropology, University of Oxford. He was an Executive Director of the Centre for Nepal Studies UK (CNSUK) between 2010 and 2011. Amongst his publications is *Nepalis in United Kingdom: An Overview* (2012). He completed his PhD at the University of Reading (2007) on social capital and community-based institutions, and holds a master's degree from the University of Goteberg, Sweden and an MBA from Tribhuvan University, Nepal. His research interests and publications include collective action, natural resource management, land and forest tenure, development and migration.

Dr Ganga Ram Dahal is a consultant for forest tenure policy to the Food and Agriculture Organization of the United Nations, Technical Cooperation Program. He is an international expert on forest policy and governance with solid experience in forest tenure and livelihoods-related research, analysis and advocacy at the national, regional and global levels. Key areas of expertise include: forest land tenure, governance, institutional structure, resource rights, community-based resource management, decentralization and devolution in forestry. Dr Dahal completed his PhD on forest governance at the University of Reading in the UK and worked at CIFOR for five years as a Policy Researcher and at Rights and Resources Initiative for five years as a Regional Facilitator for Asia. He has published books on forest policies and decentralization, and more than five dozen articles in international journals.

Dr Binod Devkota is a District Forest Officer with many years of experience working in the Department of Forests, Nepal. Binod completed his PhD research through Charles Sturt University, Australia in 2012, exploring the socio-economic outcomes of community forestry in Nepal, having earlier received a master's degree in environmental management and development from the Australian National University. His research interests include the interactions between community forestry, climate change and the livelihoods of forest-dependent people in Nepal.

Dr Robert (Bob) Fisher is an anthropologist who has worked extensively in community forestry and related fields since his work with the then-Nepal-Australia

Forestry Project between 1987 and 1989. He was Deputy Director of RECOFTC (the Regional Community Forestry Training Centre for Asia and the Pacific) from 1997–2001. His experience in community forestry research and consultancy includes Nepal, India, Thailand, Cambodia, Vietnam, Kyrgyzstan, Mongolia, Liberia, Ghana and Papua New Guinea. He was co-author with Don Gilmour of the well-known book *Villagers, Forests and Foresters: The philosophy, process and practice of community forestry in Nepal* (1991) and has since published widely on community forestry. He holds positions at the University of Sydney and the University of the Sunshine Coast, where he is engaged in a major research project on community forestry in Papua New Guinea.

Dr Popular Gentle is Program Director for Care International in Nepal, and an Adjunct Research Fellow at Charles Sturt University, Australia. Popular has over two decades of professional experience in the areas of community-based natural resource management, climate change adaptation, gender equality and a pro-poor- and rights-based focus on development. Born and raised in the western remote hills of Nepal, Popular has an in-depth understanding of underlying causes of poverty, vulnerability and social injustice. Popular obtained a master's of forestry science from University of Canterbury, New Zealand, and a PhD from the Charles Sturt University, Australia.

Dr Mohan Poudel is REDD+ Specialist in the REDD Implementation Centre, Ministry of Forests and Soil Conservation, Nepal. Growing up in a rural village in Nepal, Dr Mohan Poudel undertook his bachelor's of science degree (forestry) in Nepal, a master's degree in the Netherlands and his PhD at Charles Sturt University in Australia, where he investigated the outcomes of the implementation of REDD+ through community forestry in Nepal. Mohan has previously held positions of District Forest Officer and Planning and Monitoring Officer with the Department of Forests in Nepal. In his current position as REDD expert in the REDD Implementation Centre, Nepal, he is engaged in the development of REDD+ policy including coordination of the national forest reference level (FRL) development process and the REDD+ Himalaya project in Nepal. Mohan's key areas of interest and expertise include community forestry, climate change and REDD+.

Professor Ridish K. Pokharel is a Professor in the Institute of Forestry at Tribhuvan University and currently holds the position of Chief of the Planning Division, Tribhuvan University, Nepal. He received his PhD degree from Michigan State University, USA after investigating the outcomes of community forestry in Kaski District of Nepal, and his master's degree from Araneta University investigating social forestry programs in the Philippines. He has undertaken numerous research projects related to community-based forestry. His research interests include community-based forest management, governance, gender, and criteria and indicators.

Dr Digby Race has contributed to, and led research teams, exploring the socio-economic dimensions of rural development in the Asia-Pacific region and

Australia over the past 25 years. A major part of Digby's work during the last decade has been to improve understanding about how community-based forestry can better contribute to the livelihoods of rural communities. During the period of research covered in this publication, Digby was employed as a Senior Research Fellow at the Australian National University (ANU). He has recently joined the Tropical Forests and People Research Centre at the University of the Sunshine Coast, and has continuing affiliations with the ANU, Charles Darwin University, Charles Sturt University and the University of Gadjah Mada (Indonesia). When not working with research partners in tropical forests, Digby enjoys living with his family on a small farming property in northeast Victoria.

Dr Eak Rana has recently completed his PhD studies at Charles Sturt University in Australia, investigating the trade-offs and synergies between biodiversity, livelihoods and carbon arising from implementation of a REDD+ pilot in community forests. Prior to commencing his doctoral studies, he was the coordinator of the REDD+ project with the International Centre for Integrated Mountain Development (ICIMOD). He has more than 12 years of experiences in community forestry, livelihood and climate change. His major areas of research are the intersection of terrestrial ecosystem services, livelihoods and climate change.

Dr Krishna Shrestha is Senior Lecturer and Convenor of Masters of Development Studies at the School of Social Sciences, University of New South Wales, Australia. Prior to this, Krishna was Program Director of Urban Planning and Policy Program at the University of Sydney. In 2016, he was a Visiting Senior Research Fellow at Asia Research Institute, Singapore National University and a Visiting Scholar at Cambridge University. Krishna is an urban and environmental geographer with core interest in *socio-environmental justice*. Specific areas of his research encompass political ecology, climate change adaptation, international development planning and policy, and disaster governance, in particular the intersection of the four with his work on *socio-environmental justice*. His academic research, teaching and professional engagements (mostly in Nepal, India and Australia) are interdisciplinary, empirically grounded and impacts oriented, with a strong commitment to integrate science and policy to enhance inclusive processes and outcomes at the local community level. Krishna began his working life as a Forester with the Department of Forests in Nepal and completed his PhD on collective action and equity in Nepalese community forestry at the University of Sydney.

Dr Richard (Rik) Thwaites is a Human Geographer in the Institute for Land, Water and Society and the School of Environmental Sciences at Charles Sturt University, Australia. He has taught across the fields of ecotourism, sustainable development and environmental management, recreation and land use planning, and social research methods. Rik's PhD research explored land management practices and policies in the heavily degraded steppe grasslands of Inner Mongolia, China. His ongoing research interests are situated in the dynamic interface between local communities and their natural environments, incorporating issues associated with protected areas, climate change and local livelihoods, with particular focus on policies and interventions that seek to integrate socio-economic and environmental outcomes in less developed countries.

Dr Krishna R. Tiwari is a professor in the Institute of Forestry at Tribhuvan University and Coordinator of the NORHED-SUNREM Himalaya Project, Institute of Forestry, Nepal. He undertook his PhD studies in soil science and natural resource management at the Norwegian University of Life Sciences, Norway. His primary research interests are related to watershed management, gender, climate change, governance, and criteria and indicators.

1 Community forestry in Nepal

Origins and issues

Richard Thwaites, Robert Fisher
and Mohan Poudel

Introduction

When community forestry (CF) was first introduced in Nepal in the late 1970s, it was established in local communities that were, to a large part, isolated from outside in terms of local economy and livelihoods. Most rural people lived semi-subsistence lifestyles heavily dependent on local forests and farmlands for their livelihoods. The initial concern of CF was to provide for local management of forests that would contribute to forest conservation and reforestation, later expanding to include the link between forest resources and livelihoods. CF is now widespread in Nepal, particularly across the region of the Middle Hills, with a very high proportion of rural population involved, and is widely recognized as one of the most successful examples of community forestry in Asia. While noting these successes, the handing over of responsibility for forest management from government to communities has presented new challenges associated with local autonomy and decision-making, institutional governance, tenure and rights, as well as raised issues of equity in decision-making and benefit distribution. The establishment of local Community Forest User Groups (CFUGs) has resulted in a variety of environmental, social and economic outcomes for local communities. In many communities, CF provides a central institutional focus, bringing the community together around a common purpose, developing networks and skills, and building local expectations and capacities. In some cases, these groups have gone beyond the focus only on forest management to contribute to a broader development agenda in the local community.

Since the late 1970s and 1980s when CF emerged within the localized semi-subsistence economies of rural Nepal, the broader context of CF has changed greatly. Today, CF finds itself embedded in a more complex global economy and confronted by rapidly changing global and national pressures and priorities. Climate change and international policies such as REDD+, as well as contemporary social change, including extensive labour migration and feminization of rural agriculture, all present new challenges and opportunities for CF. The skills, structures and experiences of CF institutions present local communities with the capacity to respond to these challenges and opportunities in ways that may not have been otherwise possible. While bringing opportunities for individuals and for CF groups, these changes also place new pressures on CF and the local CF institutions, introduce new stakeholders and expectations, and could provide new challenges and threats to the operation of locally based CF institutions, especially the pressure to meet externally imposed objectives, such as carbon capture. As a program embedded in local institutions, CF requires ongoing local commitment to be sustainable and successful, yet these global 'pressures' could have implications for local autonomy and local benefits.

The purpose of this book is to explore, through a combination of review and reporting on primary research, some of the experience of CF in Nepal as a model for improving forest management and supporting local livelihoods, as well as the experience of CF in adapting to a rapidly changing world, how it has responded to the opportunities and challenges presented, and how it might move forward. As a relatively 'mature' approach to community-based natural resource management (NRM), the outcomes and implications of these responses in Nepal have relevance to community forestry and community-based approaches to natural resource management around the world that are also experiencing global pressures and opportunities. We approach the writing of this book from a critical political ecology perspective, based on empirical evidence, and whereby change is investigated not just from an environmental perspective, but from a social and political perspective that recognizes fundamental inequalities and power relationships within society.

The specific aims of this book are to:

- Provide a historical perspective on the development of community forestry in Nepal, and the experience of community forestry in delivering on fundamental environmental and social objectives;
- Explore the implications of the dynamic global context of community forestry in Nepal, and the responses of community forestry institutions to these global influences; and
- While providing specific information from Nepal, identify issues and responses or strategies that have global relevance and application.

People and forests

It has been well established that forests are crucial to the maintenance of life on earth, containing over 75 percent of the world's terrestrial biodiversity (FAO 2016), providing protection for soil and water resources, providing a source of timber and a range of other products that support life and livelihoods for forest-dwelling people as well as people living beyond the forests. As the 'lungs of the planet', forests also filter pollution from the atmosphere, provide a source of oxygen and a sink for carbon, thus making an important contribution to achieving a balance in the atmosphere that makes the planet amenable to life as we know it. Forests also have cultural and spiritual value for human populations, as well as providing recreational opportunities. But in addition to providing a wide range of services, forest ecosystems play an important economic and livelihood role, providing direct employment to over 10 million people worldwide (FAO 2010). Further, Munang et al. (2011) report that about 410 million people are highly dependent on forests for subsistence needs as well as income and the livelihoods of 1.6 billion people depend to some extent on forest goods. So, forests provide many benefits essential to life on Earth, as well as supporting the quality of human life.

It follows that any loss or degradation of forests will have many consequences for ecosystem function and biodiversity, and for delivery of services at the global level such as carbon regulation associated with global climate change, at the regional level such as regulating temperature and humidity to modify climate, and at the local level such as flood mitigation, water supply, and provision of livelihood resources and tradeable goods. Loss or degradation of forests could have dramatic consequences for the lives and livelihoods of forest-dependent peoples, as could expansion or improvement of forests.

While there are many factors causing forest degradation and loss – including anthropogenic fire, unsustainable management, over-grazing, urban expansion and development of infrastructure – FAO (2016) has identified the most significant direct factor as conversion of forests for agricultural use (FAO 2016). This can occur with the expansion of industrial agriculture, but is also associated with small-scale subsistence agriculture. FAO (2016) reports that 5,000 years of human activity converting forests to other land uses has resulted in a loss of 1.8 billion hectares (ha), a decline of almost 50 percent in forest area. In addition to the direct factors, there are a number of underlying factors that may cause forest degradation and loss. These include perverse incentives associated with lack of or insecure tenure and lack of recognition of subsistence and public good benefits from services that sustainable management of forests would deliver and that result in the liquidation of forest assets to realize immediate market value for short-term gain over these longer term subsistence and public good benefits (Barbier and Burgess 2001; Munang et al. 2011). In recent decades the greatest net loss of forests has occurred in tropical and low-income countries where populations are growing. Between 2000 and 2010, FAO (2016) report an annual net loss of tropical forest of 7 million ha, almost matched by an annual net gain in the same regions in agricultural land of 6 million ha. In the tropics and sub-tropics, large-scale commercial agriculture accounts for about 40 percent of deforestation, with local subsistence agriculture accounting for 33 percent and the remainder a result of infrastructure development, urban expansion and mining.

Keenan et al. (2015) reported that in 2015, 3,999 million hectares (ha) of land was covered by forest, equivalent to 31 percent of global land area, with a further 1,204 million ha designated as 'other wooded land'. The area of forest was down from 4,128 million ha in 1990, a net decline of over 3 percent, though the rate of net forest loss had more than halved over this 25-year period from a loss of 7.3 million ha per year from 1990 to 2000, to 3.3 million ha per year between 2010 and 2015. However, over the 25-year period from 1990 to 2015, the area of natural forest declined by 6 percent from 3,961 million ha to 3,721 million hectares, offset by an increase of planted forests of 66 percent from 168 million ha to 278 million ha.

The sustainability of forest use and management, forest condition and local livelihoods are closely connected. The challenge of feeding the growing world population is made more difficult by ongoing land degradation, water and land scarcity and climate change (FAO 2016). The FAO (2016) points out that as well as supporting local livelihoods, forests are fundamental for food security, harbour over 75 percent of the world's terrestrial biodiversity and contribute to climate change mitigation. These diverse benefits are also recognized by the Sustainable Development Goals (SDGs) (UN 2015), agreed to at the United Nations Summit on Sustainable Development in 2015, which identify 17 'universal and transformative' goals which seek to achieve sustainable development in a balanced and integrated manner. SDG 15 is explicitly relevant to the management of forests: "protect, restore and promote sustainable use of terrestrial ecosystems, sustainably manage forests, combat desertification and halt and reverse land degradation and halt biodiversity loss". Other goals are also relevant to the roles that forests play: SDG 1, ending poverty; SDG 2, achieving food security and promoting sustainable agriculture; SDG 6, availability and management of water; SDG 7, access to energy; SDG 13, combating climate change.

While the 17 identified goals are described as being 'integrated and indivisible', the great challenge for forest management will be to contribute to ending hunger, achieving food security and promoting sustainable agriculture (SDG 2), without compromising the protection of forest ecosystems and biodiversity promoted in SDG 15, or contributions to water supply, energy, poverty alleviation and climate mitigation (FAO 2016).

Community-based forest management and community forestry

Community-based approaches to forest management started around the world in response to recognized failures in government management of forests. Government management of forests was often found to be inefficient, ineffective and corrupt, as well as resulting in widespread loss of forests, forest degradation and reduced forest productivity.

Much of the global loss and degradation of forests outlined above has occurred under formal government management of forests, and there has been a general understanding that government management has not been successful in halting deforestation. There is also recognition that loss of forests under government management regimes has placed great pressure on the livelihoods of those people who depend on the forests. Numerous authors have concluded that a government-centred forest management approach cannot achieve sustainable forest management and poverty reduction amongst rural communities. In response, projects around the world have introduced more participatory approaches of community-based forestry (CBF), in which local communities that have a stake in the forests act collectively in the management and extraction of resources from local forests. Such participatory or collective approaches have been implemented in many different ways, with different tenure regimes applied, different authorities given to the local community, and different control over decisions and returns to community and government. In his review of 40 years of community-based forestry, Gilmour (2016, p. 2) defines community-based forestry as including "initiatives, sciences, policies, institutions and processes that are intended to increase the role of local people in governing and managing forest resources". Gilmour (2016) reports a substantial increase in the area of forest land under different community-based forestry regimes over the past two decades, noting that the transfer of power to local people is based on the "explicit assumption that the transfer of rights to communities will lead to SFM (sustainable forest management) and improvement in key environmental, social and economic outcomes that benefit smallholders and communities (RECOFTC 2013)" (Gilmour 2016, p. 3).

Given the very different local contexts and histories, CBF regimes are also very diverse, particularly in relation to the tenure regimes and rights granted to local communities over use and management of forests and their level of participation in decision-making. Gilmour (2016) has described a generic spectrum of CBF, in order of increasing strength of rights, participation and empowerment:

- participatory conservation,
- joint forest management,
- community forestry with limited devolution, and
- community forestry with full devolution.[1]

At this point, we note that there is potential confusion around the use of terminology which has evolved over the years, from social forestry to community forestry, to community-based forestry. Today, 'community-based forestry' has become the generic term as presented by Gilmour (2016), and community forestry falls within this generic term. Individual countries have adopted different names for their programs that all fit under the CBF term. Community forestry (CF) has been adopted as a term to describe approaches in different countries, but the term CF describes quite different approaches in different countries. In this book, we use the term community forestry to describe the specific program as it has been designed and implemented in Nepal, as discussed below.

Why look at Nepal?

Nepal has amongst the longest history of implementing community-based forestry in Asia (adopting the specific term 'community forestry' to describe its first program in this area), and was one of the earliest programs anywhere in the world. Nepal's community forestry program is one of the most comprehensive programs in terms of its reach, number of people involved, number of groups and geographic area covered as a proportion of national land area. Community forestry in Nepal is also one of the more radical programs in terms of the transfer of rights to the local community, including offering people the right to form user groups and to manage community forest under specified conditions, if they are willing and able to do so. In other programs in Asia, formation of community forestry groups is at the discretion of government. The legislation in Nepal allows for considerable local control through devolution of substantial rights such as for decision-making and distribution of benefits from forest management to the local community. This devolution is more extensive than programs in other countries, such as joint forest management in India, where control is explicitly retained in the Forest Department. Thus, the extent of benefits to communities is potentially greater in Nepal than in most other communities. So, Nepal provides a model of what is possible, and what the issues and challenges might be of implementing this more 'democratic' approach.

Further, community forestry in Nepal has been well documented and promoted as a model, and is well known amongst academics and practitioners around the world. A number of countries have attempted to adopt some aspects of it including its terminology and structure; for example a number of countries have adopted the concept of the local forest user group. Certainly in Asia and even in Africa there has been much transfer of concepts from Nepal – frequently people writing in academic journals about other countries refer to reports/publications from Nepal in a comparative sense. Nepal could be described as a living laboratory for the implementation of community forestry and testing of theories of democratization and community-based NRM. In this context, it is relevant to take a critical look at what is happening, good and bad, and how it is changing in Nepal, and how that might inform people's thoughts in Nepal and elsewhere. The reality of devolution of rights to local communities, equity in local decision-making processes, design and governance of local institutions, and related issues are themes that will be explored throughout the book.

Characteristics of Nepal – population, forests and forest dependence

Nepal is a landlocked country that sits between India and China on the southern slopes of the Himalaya range, with an area of 147,181 square kilometres. Five physiographic zones have been described (LRMP 1986), rising from the Terai on the plains along the Indian border, to the Siwalik or Inner Terai in the low Churia Hills, then to the Middle Mountains, the High Mountains, and finally in the north along the Chinese border, the High Himal. These zones vary in their topography, soils, climate and forest types, as well as their population densities, ethnicities, land uses and livelihoods. Elevation rises from about 60 metres in the south, to the highest Himalayan peaks above 8,000 metres in the north, and correspondingly, climate varies greatly from sub-tropical humid, to a dry and cold alpine climate. Being found across the juncture of the eastern and western Himalayan regions and from sub-tropical to alpine conditions, Nepal has a very rich biodiversity with enormous variation in habitats and species mixes across both east-west and altitudinal north-south lines. Forests in the south are predominantly Sal forests or mixed broadleaf forests, transitioning into higher proportions of coniferous species with higher altitude. In the Middle Mountains, forests are very diverse, containing mixed hardwoods, pines and oaks, but in the higher mountains fir and pine predominate with rhododendron and then above the tree line in the High Himal, tundra shrubs and grasslands are predominant.

Many studies of Nepal have adopted a simpler system of three 'ecological zones' or regions: the Terai, the Middle Hills or Hills, and the Mountains (Figure 1.1), which present clearly different climatic conditions, as outlined in Table 1.1, and have often been used for administrative and statistical purposes. This book will follow the preference of national statistical assessments, such as represented by the Central Bureau of Statistics in the census data and in the National Living Standards Surveys, and the Department of Forests' Community Forestry Division by referring to these three ecological zones for the purpose of spatial considera-tion of data. It should be noted that the three ecological zones follow the political boundaries of the districts, not the topographic or true ecological boundaries. The

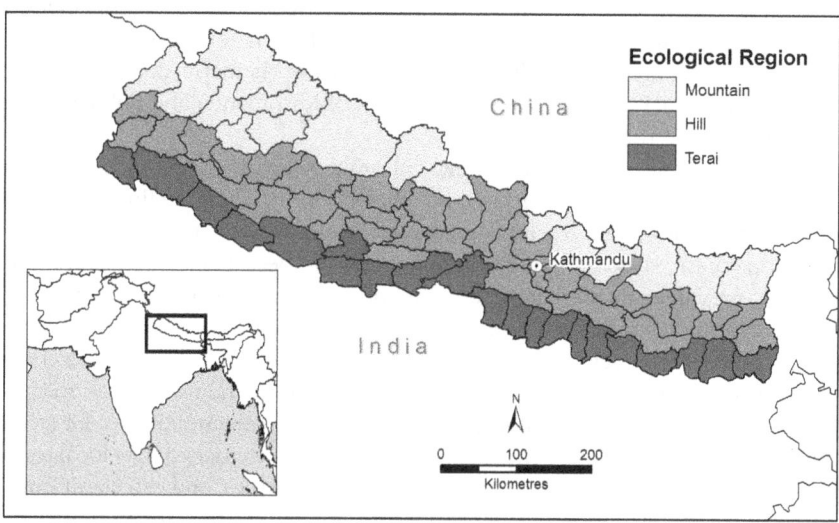

Figure 1.1 Map of ecological zones of Nepal showing district boundaries
(Source: adapted from NPC 2014)

Table 1.1 Climatic conditions of Nepal across ecological zones

Physiographic zone	Ecological zone	Climate	Average annual precipitation	Mean annual temperature
High Himal High Mountain	Mountain	Arctic/Alpine	Snow/rain 150–200 mm	<3°C–10°C
Middle Mountain	Hill	Cool/warm	275–2,300 mm	10°C–20°C
Siwalik Terai	Terai	Tropical/sub-tropical	1,100–3,000 mm	20°C–30°C

(Source: WECS 2005)

result is that separation of data for statistical purposes by ecological zones does not always reflect physiographic reality. For example, Sindhu Palchok District, which was the focus of early international development projects focused on community management of forests and has a high number of Community Forest User Groups (CFUGs), is designated as a High Mountains district, yet physiographically, it contains both High Mountain and Middle Hills environments. While recognizing the potential confusion when some ecosystems within a Mountains district would more naturally be described as being of Middle Hills ecology, we adopt throughout this book the terminology used elsewhere by calling these 'ecological' zones.

Estimates of population vary dramatically, but at the last official census in 2011, the population of Nepal was just under 26.5 million people living in 5.4 million households, with a growth rate of 1.35 percent per annum since the 2001 census (down from an annual growth rate of 2.25 percent over the preceding 10-year period 1991 to 2001) (CBS 2014). Distribution of the population is uneven across the country, with 50.3 percent living in the Terai, 43 percent in the Hills, and only 6.7 percent in the Mountains. Population density is relatively low in the Mountains with only 34 people per square kilometre, compared to 186 in the Hills (which contain the urban centres of Kathmandu and Pokhara), and highest at 392 in the Terai.

Nepal is a multi-religious and multi-cultural society, with 125 caste and ethnic groups speaking 123 languages and dialects (CBS 2014). Hinduism is the dominant religion, and despite considerable social change in recent decades, a traditional caste structure and social hierarchy has been maintained and is particularly prevalent in rural areas. At the bottom of the social hierarchy are the Dalit, or 'untouchables', who remain socially and culturally oppressed. UNDP (2009) identified seven sources of inequality and exclusion, based on gender, caste, ethnicity, language, religion, isolation from Kathmandu, and location (hills or plains) that have been reinforced over time by government policies. Discrimination based on caste, ethnicity and gender remains entrenched into local social systems, particularly in rural areas, dictating opportunities, and can significantly obstruct participatory involvement of some groups in local institutional and social interactions.

The 2011 census showed that 83 percent of the population lives in rural areas (CBS 2014). According to the National Living Standards Survey (CBS 2011a), 76 percent of all households are agricultural households, and more than half of these have less than 0.5 ha of land to farm, and only 4 percent have more than 2 ha. In 2011, agriculture provided the main source of self-employment for 61.3 percent of the employed population (though this was down from 70.7 percent in 1996), as well as 2.8 percent of wage employment (though this too was down from 12.2 percent in 1996). Substantial differences can be seen in Table 1.2, with higher dependence on agriculture for employment amongst women, in rural areas and in the mountains.

Agriculture remains the mainstay of rural economies and livelihoods, and farming is mostly small scale and subsistence based, though there would appear to have been a reduction in the number of rural people depending on wages from

Table 1.2 Proportion of employed individuals by employment in agriculture 2011 (percent)

	Gender		Urban/rural		Ecological zone		
	Male	Female	Urban	Rural	Mountain	Hill	Terai
Wage agric	2.7	2.8	1.0	3.1	1.6	1.5	1.2
Self agric	53.6	67.7	30.5	67.4	70.8	62.8	58.3

(Source: CBS 2011a) (wage agric = wage employment in agriculture; self agric = self-employed in agriculture)

agricultural labour in recent years. In 2013, agriculture contributed 34.7 percent of GDP, compared to 15.0 percent from industry and 50.3 percent from services (NPC 2014), though how well this reflects the value of subsistence farming to rural households is unclear and it does not reflect the disparity of economic development between urban and rural areas, or the heavy dependence on agriculture outside the cities.

Despite the growing urbanization of Nepal, much of the population is still heavily dependent on forests for resources. As an example, 64 percent of all households in Nepal still depend on firewood as their primary source of cooking fuel, with bottled LP gas being the primary source for 21 percent and cow dung for 10.4 percent of households. There is a stark geographic difference in fuel used, with 73.1 percent of rural households depending primarily on firewood, and while 67.7 percent of urban households use LP gas, over one-quarter of urban households still use firewood as their primary fuel (CBS 2014). In the Kathmandu Valley, 93 percent of households depend primarily on LP gas (CBS 2011a). Dependence on forests does not only relate to fuelwood, however, with rural communities relying on a range of forest species for livestock fodder and bedding, manure for their fields and for other needs such as building materials, food and medicinal plants. According to the national Forest Resource Assessment undertaken from 2010 to 2014, over 40 percent of the country's land area is covered by forest (5.96 million hectares), and if shrublands and other wooded lands (less than 10 percent canopy) are included, then the cover is almost 45 percent of the country (DFRS 2015). While this raw figure for forest area seems considerable, heavy dependence on forest resources suggests that changes in area and condition could have dramatic consequences (positive or negative) at the local level.

The continued dependence on forest resources for firewood and as part of the agricultural and pastoral systems, as well as the vulnerability of Nepal's economic development to shocks, can be seen by the recent experience of the blockade of the Indian border in late 2015 and early 2016. Being a landlocked country, Nepal is heavily dependent on imports coming through India. The blockade (imposed by Madheshi political groups from Nepal's southern border regions with the apparent support of the Indian authorities, in response to the passing of the new constitution of Nepal) stopped import of many products, particularly fossil fuels, which presented a particular challenge in the aftermath of the

2015 earthquake in central Nepal. During the blockade, cooking gas became very scarce and expensive, and prices for a bottle of gas in Kathmandu rose from around 400 rupees per bottle to around 6,000 rupees. The result was a dramatic increase in the dependence on firewood for cooking, by households as well as commercial premises. According to Tiwari (2015), the increased demand for firewood resulted in the harvesting of more trees, including illegal harvesting. Tiwari further reports on studies that indicate that income shock resulting from a combination of the earthquake and the blockade of the Indian border would push an additional 1.5 million people below the poverty line, and that as the poor rely more heavily on common property resources than others, this will only place more pressure on forests and forest resources. This is a good example of how vulnerability and dependence on forest resources may be influenced by political events.

Income and poverty

Nepal is a poor country and is designated as one of the world's 'Least Developed Countries' being "characterized by constraints such as low per capita income, low level of human development, and economic and structural handicaps to growth that limit resilience to vulnerabilities" (NPC 2014, p. 1). The National Planning Commission describes the economy as "characterized by a low level of economic growth, subsistence agriculture, ever widening trade deficit, and low levels of domestic savings" (NPC 2014). The Central Bureau of Statistics put the nominal GDP per person in 2014 at USD703, or under USD2 per day (CBS 2014), with an annual GDP growth rate of 3.04 percent. The Human Development Index (HDI) goes beyond a measure of development based on a single indicator of income, being a composite of three specific dimensions considered to represent people's opportunities and choices (life expectancy at birth, knowledge measured as years of schooling and adult literacy, and standard of living as GNI per capita). The most recent figures show Nepal is in the 'low human development' category, ranked 145 out of 188 countries on the HDI (UNDP 2016), though Table 1.3 (from UNDP 2016) shows a rapid improvement in indicators and HDI values over recent decades.

Measurements of human development indicators in Nepal reveal inequalities across ecological zones and social groups (UNDP 2014). Based on data from

Table 1.3 Variation in Nepal's HDI indicators over time

	Life expectancy at birth	Mean years of schooling	GNI per capita (2011 PP$)	HDI value
1980	46.6	0.6	961	0.279
2000	62.4	2.4	1,563	0.451
2014	69.6	3.3	2,311	0.548

(Source: UNDP 2016)

2011, the Human Development Index values are considerably higher for urban than rural areas, and also higher in the Hills than in the Mountains or Terai (though this may be influenced by the presence of major urban centres in the Hills region) (Table 1.4). Considerable differences can also be seen in the individual indicators, particularly adult literacy, mean years of schooling and per capita income, which is well over double the value in urban areas compared to rural areas. Annual income-based economic growth in the agriculture industry, which is largely rural based, is only 1.3 percent, while in the service industry it is 6.0 percent (NPC 2014), highlighting growing inequalities between urban and rural dwellers.

While still predominantly a rural nation, Nepal is rapidly urbanizing. A combination of better employment and education opportunities in the cities, as well as the Maoist insurgency (1996–2006) and most recently the earthquakes of April and May 2015 have seen many people leaving rural areas for urban life. Based on the census figures, the population of Kathmandu District (not the same as Kathmandu City, or of Kathmandu Valley) has increased progressively by about 60 percent each decade from 1981 to 2011 (CBS 2013) (Table 1.5), compared to the national population increase of 23 percent, 25 percent and down to 14 percent in the most recent decade. As well as this rural-urban migration, there is also a trend in migration to international locations. Much of the migration in recent years has been by working age males, seeking labouring work in the city, or in India or further afield (see Chapter 9). This trend can be seen through the gender ratios derived from the 2011 census data (CBS 2014). In urban areas, there were

Table 1.4 Spatial variation in Nepal's HDI indicators in 2011

	Life expectancy (years)	Adult literacy (percent)	Mean yrs schooling (years)	Per capita income (PPP USD)	HDI
Nepal	68.8	59.57	3.9	1160	0.541
Urban	68.93	79.27	4.94	2248	0.630
Rural	68.81	54.98	3.69	936	0.517
Mountain	66.98	51.46	3.02	965	0.496
Hills	69.02	66.77	4.46	1316	0.569
Terai	68.85	54.24	3.52	1052	0.521

(Source: UNDP 2014 Annex 4)

Table 1.5 Population of Nepal and Kathmandu District by census 1981, 1991, 2001, 2011

		1981	1991	2001	2011
Nepal	Population	15,022,839	18,491,097	23,151,423	26,494,504
	Percent change		+23	+25	+14
Kathmandu District	Population	422,237	675,341	1,081,845	1,744,240
	Percent change		+60	+60	+61

(Source: CBS 2013)

1.04 males for every female. In rural areas, the ratio was much lower, with only 0.92 males per female. For the whole country, the figure was 0.94, reflecting the migration of males out of the country. Overall, 53 percent of households in Nepal have at least one absentee member, and 33 percent of households have at least one family member living overseas, though there are various reasons described for living away, including for family reasons, study or work (CBS 2011a).

As well as a demographic change, these patterns of migration have brought economic change to rural areas. While it may have been the powerful and better educated who left rural areas during the Maoist insurgency (Rechlin et al. 2007; Nightingale and Sharma 2014), leaving poorer farming families behind, the recent labour migration has seen many members of poorer families, mostly males, leaving their rural villages. The resultant remittances now make an important contribution to household finances. Between 1996 and 2011, the proportion of total household income that comes from farming has dropped from 61 percent to 27.7 percent, while share from other sources including remittances has risen from 16 percent to 35 percent (CBS 2011b). The number of households receiving remittances more than doubled over this 15-year period, from 23 percent of all households to 56 percent. The total remittance received in 2011 was 259 billion Nepal rupees (approximately USD2.5 billion), and about 80 percent of this came from international sources (up from 55 percent in 1996) (CBS 2011b). The World Bank estimates that, in 2015, international remittances made up 31.8 percent of national GDP of Nepal, a dramatic rise from an estimated 2.4 percent in 2001 (World Bank 2016).

Forests in Nepal and their management

As has been seen above, the majority of Nepal's population is rural and many rural people rely on semi-subsistence agriculture (crops and livestock) for their livelihoods. Forests also provide important resources for livelihoods and incomes of rural and urban people across Nepal. Hill (1999) reported that almost 90 percent of people in Nepal depend on forests for fodder, fuelwood, food, building materials, medicinal plants and fertilizers for some contribution to their subsistence needs. While urbanization and migration have changed local economies since then, forests remain an integral part of the daily lives of rural people in Nepal, contributing to the rural economy in many ways. Forests provide employment and value through the harvesting, processing and marketing of forest products: in 2005, over 100,000 people were directly employed in forest management and harvesting and delivering government services (though this was about 45,000 fewer than in 1990, largely as a result of the Maoist insurgency which made life in rural areas dangerous for government employees) (FAO 2014). This figure excludes all employment through community-based forest management and all gathering of forest products for personal use. Within local communities, forests are critical to agriculture and livelihoods, providing employment, clear water, construction materials, food, medical supplies, firewood for cooking and heating, forage, fodder and bedding for livestock; they also provide the primary source of nutrients as compost for agricultural fields.

Forests provide opportunities and resources for which local people who would have no viable alternative options without the presence of local forests, and deforestation could have significant consequences for local livelihoods and economies. Deforestation (including degradation of forest to lower productivity vegetation with fewer trees, and conversion of forest to other uses such as agriculture) has been a serious problem in Nepal (Bhattarai, Conway and Yousef 2009), especially in the Terai, where most forests are still government managed. Management can be seen to greatly influence local people's interactions with forests, their use of forest resources and the outcomes for forest condition.

Forest management regimes have changed greatly over the last hundred years. Up until 1957, forest was managed under the Rana regime as an essentially feudal system of central government control and granted to favoured individuals, frequently with the objective of converting forest to agricultural land and profiting from harvesting of timber (Gautam, Shivakoti and Webb 2004; see also Chapters 2 and 6 in this volume). Private (feudal) forests in Nepal were nationalized in 1957 through central laws that not only took over control of private forests, but also, according to some views, over-rode traditional management systems. It is often stated that this nationalization of private forests led to deforestation and degradation (e.g. Gautam, Shivakoti and Webb 2004), but an alternative view presented in Chapter 2 (this volume) is that the chaos that ensued from the fall of the Rana regime in 1951 was the main driver causing this process of forest loss and degradation, and that forest loss began earlier than the nationalization in 1957. Regardless of the cause, the experience of forest loss and degradation and the failure of the government system of the time to address the problem presented the conditions for the introduction of a more participatory approach to forest management, ultimately delivering the evolution of community-based forest management approaches, the intention of which was to turn this ongoing process of forest loss and degradation around. (Chapter 3 in this volume and Table 3.1 provide a more detailed discussion of broad trends in forest and shrubland area, including the gradual decline in area over a number of decades and the apparent improvement in both forest area and conditions in the most recent decade.)

Throughout these changes in central government policy on forest management, local communities continued to demand access to forests and to manage them wherever possible according to their own traditions and culture, as well as based on collective use rights (Dev and Adhikari 2007). Forests have been, and remain, an integral part of the farming system which sustains the rural households and economy of Nepal (Kanel 2006). Despite the exclusion of local people from management by the Department of Forests throughout the 1960s and 1970s, local households continued to seek access to needed products, and their traditional history of management and use of the forest provided an environment ripe for the establishment of a more participatory approach. Further, the existence of indigenous informal systems of forest management, often formed in the 1960s and 1970s as a response to local perceptions that there was a vacuum in forest management, adds evidence for the potential of a participatory management approach (see Chapter 2). The National Forest Plan of 1976 recognized the need for local participation and set the objective of local involvement and partnership in the protection and utilization of forests (Kanel 2006).

According to Mahat, Griffin and Shepherd (1986), the First Amendment of the Forest Act 1977 identified six categories of forest, some of which enabled the involvement of local people and communities in the conservation, development and management of forests. The different social and ecological contexts of forest use and development around Nepal resulted in a number of different models of community-based forest management being established over the years, the first of these to emerge being community forestry. Ojha (2014) has identified four different CBF regimes that have been supported by different international donors, managed by different government departments or units, under different themes or policy goals, and employing diverse strategies for service delivery. These are: community forestry (CF), community-based leasehold forestry, collaborative forest management (through partnership between local communities and the government Department of Forests), and community-based buffer zone management around protected areas. Of these, CF has been around the longest, since the late 1970s, is the most widespread and involves the greatest number of people and forest area. The later regimes were established in part to respond to perceived weaknesses or gaps in the achievements of CF, such as the establishment of collaborative forest management (CFM), which provides a stronger role for government in the management 'collaboration' and was initially designed to return 75 percent of commercial income from timber to government. CFM was focused in part on the fertile plains of the Terai, where government was reluctant to hand over high-value Sal forest to community management under the CF regime (Ojha 2014).

This book adopts the terms community-based forestry (CBF) as a generic term that describes a participatory approach to forest management that provides for devolution and decentralization of authority and responsibility to local communities for collective management of forests as a common pool resource. The main focus of the book, however, is on community forestry (CF), which is one of the abovementioned forest management regimes with very specific meaning.

In Nepal, a community forest (as defined by the Forest Act 1993) is a part of the national forest that has been handed over to a local user group for its development, conservation and utilization for the collective interest. That is, the land remains owned by the government, but is handed over to members of the local community in the form of a Community Forest User Group (CFUG) to manage the forest for their collective benefit, with use rights subject to approval of an operational plan by the Department of Forests.

As of January 2017, 1.81 million hectares is managed as community forest, which is 30 percent of all the forest in Nepal, or over 12 percent of the area of Nepal (DoF 2017). This forest has been handed over to 19,361 local institutions, the Community Forest User Groups responsible for developing their own

constitutions and operational plans for the forests subject to approval of an operational plan by the Department of Forests, with almost 2.5 million households
involved in this process, representing about 44 percent of all the population of
Nepal, or 55 percent of the rural population.

While the number of CFUGs continues to grow year by year, their distribution
across Nepal is not even, reflecting different government priorities in different
regions of Nepal (Table 1.6). The majority of the CFUGs established to date
have been in the Middle Hills region, where mixed species forests were heavily
degraded and where there is a medium population density of 186 persons per
square kilometre (CBS 2014). While new CFUGs are still being established,
the growth is slow, as most of the national forest that is accessible and close
to local communities has already been handed over (Kanel and Kandel 2004).
More distant forests remain under government management. In the Mountains
region, there are fewer human settlements and a lower population density, and
fewer CFUGs have been established. In the Terai, population density is high, but
government has restricted the handing over of forests to CFUGs (Kanel 2004),
preferring in many cases to adopt the collaborative forest management model,
which, as described above, provides government with greater control as well as
a commercial return from the commercially valuable Sal forests. As a result, the
number of CFUGs established is quite low. While the average area of community
forests per CFUG is much higher in the Terai, the CF area per household is highest in the Mountains and lowest in the Terai.

With so many members, and with responsibility for such a large proportion of
the land and forest area of Nepal, CFUGs have become one of the largest and
most influential community-based institutions in Nepal. This provides opportunities for people across Nepal to engage in an institution designed for community
collaboration and capable of providing skills, experience and training beyond
what they might otherwise gain in their daily lives. Through membership of
CFUG Executive Committees, over 210,000 people currently have access to such

Table 1.6 CFUG database summary, January 2017

	High Mtn	Mid Hills	Terai	TOTAL
No. CFUGs	3,030	13,808	2,523	**19,361**
Total Area CF (Ha)	282,867	1,162,660	367,949	**1,813,478**
No. households (HH)	315,309	1,514,676	631,564	**2,461,549**
Mean Area/CFUG (Ha)	93.4	84.2	145.8	**93.7**
Forest area/HH (Ha)	0.90	0.77	0.58	**0.74**
Total no. cttee members	34,328	147,445	28,377	**210,150**
Mean no. cttee/CFUG	11.3	10.7	11.2	**10.9**
Female cttee members	9,776	48,308	11,317	**69,401**
Fem % cttee membership	28.5	32.8	39.9	**33.0**
No. female only cttees	162	648	267	**1077**
Fem cttees % of all CFUGs	5.3	4.7	10.6	**5.6**

(Source: based on DoF 2017)

experiences of decision-making, and many more would have experienced these opportunities over the past two decades. In a country where gender discrimination is often systemic, the potential for women's participation is also substantial, as can be seen by the data in Table 1.6 for female membership of committees, as well as by the number of CFUGs whose Executive Committees consist entirely of women.

With such an extraordinary reach across the rural communities of Nepal, covering much of the country and touching much of the population, CFUGs also provide a powerful opportunity for communication, information dissemination and training. For government, CFUGs often provide the preferred institutional mechanism for influencing local communities, and the local means of communication and information dissemination. CFUGs depend on collective action, and thus there is a strong potential for CFUGs to be much more than a local institution for managing forest resources. While initially a government program, due to the sheer number of local institutions established and proportion of people working together around common objectives, facilitation of organizations such as FECOFUN and the changes generated as a result both in forests and in rural communities, CF could be described as a national social movement.

Organization of the book

The following chapters explore many of the themes raised in this introductory discussion, considering questions associated with the history, implementation, outcomes, changing context and future directions of community forestry in Nepal. The following chapters draw on the extensive literature on community forestry in Nepal, to give some indication of current understandings, debates and questions. But they also go further, providing analysis based on the wide experience of the authors in developing, implementing and researching the challenges and outcomes of community forestry in Nepal. Some chapters present results from detailed case study research undertaken with local CFUGs. To assist the reader in understanding the physiographic and geographic context of the main case studies, Figure 1.2 illustrates the boundaries of the 75 historical districts of Nepal, highlighting those districts within which reported case studies have been undertaken. It should be noted that at the time of going to publication, arrangements for administration across the regions of Nepal are changing. The new constitution of Nepal (2015) has declared the establishment of seven new states or provinces, based largely around groupings of existing districts (though the final number and coverage of these states remains highly uncertain). The shift of power from the central to the regions under a 'federalist' system is proving difficult and remains under complex negotiation. Under the new constitution the district-level administration will have considerably reduced powers and will mainly be responsible for service delivery, with considerable devolution of decision-making power to a new form of local government that will replace the previous Village Development Committees (VDCs). However, at the time of undertaking the research and of writing this book, CFUGs were operating under

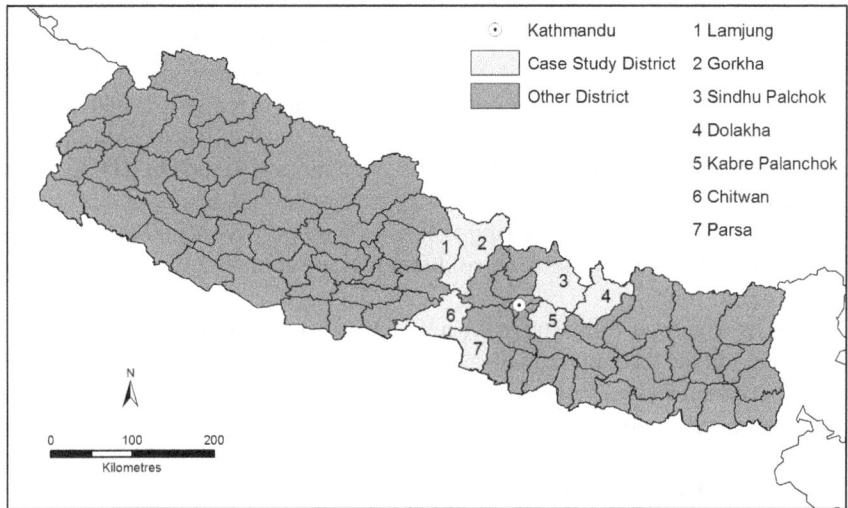

Figure 1.2 Map of districts of Nepal highlighting case study locations
(Source: created by the authors)

district-level administrative structures under the authority of the District Forest Offices, which provide technical support and to which CFUGs are responsible for review and acceptance of forest operational plans. As the districts were the operational administrative unit at the time of the research, and remained so throughout the period of writing, the chapters refer to particular districts when locating case studies, as highlighted in Figure 1.2.

Following this introductory chapter, we continue our journey in Chapter 2 by considering the history and context of the development of community forestry in Nepal, from the need for a shift away from government management of forests to establishment of the first experiments in local community involvement through creation of Panchayat Forests and Panchayat Protected Forests, and the later shift to management by groups of forest users and how these shifts run parallel to political change occurring in Nepal at the time. In this chapter, Fisher, Thwaites and Poudel explore not only the drivers of these changes, but also some of the tensions around this shift in control and the competing objectives held for managing forests. The chapter emphasizes the extent to which control of community forestry has always been contested and the continual process of adaptation to changing circumstances.

The book then moves to exploring the experience of implementation of community forestry in Nepal in delivering expected outcomes. Pokharel, Tiwari and Thwaites consider in Chapter 3 how CF has delivered on the expectation of improved environmental outcomes. They review the environmental objectives of CF and then, drawing from the literature including detailed case studies as well as national-level forest assessments and inventories, report on the outcomes against some key environmental indicators such as forest cover, density and condition, regeneration and biodiversity. Chapter 4, by Devkota, Thwaites and Race, builds on this theme by considering the livelihood outcomes for local communities from community forestry. Based on three detailed case studies, they investigate the access to and distribution of forest products and collection and mobilization of community funds in CFUGs, with a particular emphasis on institutional governance and equity of access and distribution of benefits, suggesting that achieving positive social or well-being outcomes for local people is more difficult than achieving improvements in forest conditions.

The same authors follow up in Chapter 5 with an exploration of the role of CF in delivering local community development. There is an expectation, enshrined in the community forestry guidelines, that CF will deliver community development outcomes, and it is on this basis that Chapter 5 investigates the financial investments made by CFUGs in income generating activities and in local infrastructure, as well as investments in social change through facilitating participation, building capacity, and forming social networks and community trust. While there are some positive outcomes reported, the chapter highlights institutional challenges, finding that achieving equitable outcomes remains a challenge as the most marginalized in the community remain excluded from most benefits of CF and the community development process.

In Chapter 6, Dahal, Adhikari and Thwaites consider matters of tenure and how the bundle of rights allocated to communities' influences outcomes of the transfer of those rights in terms of livelihoods and forest management. The chapter presents some of the different models of community-based forestry introduced in Nepal, and considers how the bundle of rights allocated in each influences the outcomes for local people and the forest. These experiences of forest tenure reform in Nepal are then placed in the context of reform unfolding at the global and regional levels, as well as of how Nepal's policies both inform and are informed by these global trends.

The book changes gear a little in the following chapters by considering matters of global change, how these are experienced at the local level, and how local people and community forestry are responding to these changing external conditions. Chapters 7 and 8 both consider implications of climate change and aspects of community forestry responses to climate change. Recognizing that impacts of climate change are being felt across rural Nepal, Gentle and Thwaites examine the role and potential of community forestry institutions in supporting the poorest and most vulnerable members of local communities for climate change adaptation. Based on detailed case study research, this chapter identifies fundamental inequities in how community forestry institutions deliver adaptation support, as well as highlights that improved governance and enhanced capacity, knowledge and skills are critical factors to optimize the role of CFUGs in climate change adaptation.

Community forestry has been recognized as offering great potential in contributing to global climate change mitigation efforts through carbon sequestration under the REDD+ policy initiated by the United Nations Framework Convention on Climate Change (UNFCCC). In Chapter 8, Poudel, Rana and Thwaites draw upon detailed case study research in a number of CFUGs involved in a pilot REDD+ program to explore the question of whether community forestry can continue to achieve core objectives of forest management to deliver livelihood improvement in an equitable manner while also contributing to climate mitigation goals. The chapter shows how the implementation of REDD+ with its emphasis on carbon enhancement influences CFUG management of forests towards protection rather than utilization, resulting in stricter access and forest harvest regulations, with the further result that poor and marginalized forest users are deprived of meeting their basic resource requirements. The chapter also finds that the implementation of the global REDD+ policy through locally based community forestry institutions may result in the destabilization of customary CF management systems and reduction in member commitment to community forestry.

Apart from climate change, there has been an enormous amount of change since CF was first introduced to engage relatively isolated forest-dependent communities in the management of forest resources. In Chapter 9, Shrestha and Fisher explore the complex changes to the context within which community forestry operates arising from the large numbers of Nepalis who have migrated internationally for work and whose remittances have become an important source of income for many families. The remittance economy has led to complex social changes, including the feminization of agriculture, which refers to the increasing engagement of women in agricultural work previously carried out by men. Community forestry has been affected as engagement in CFUGs has decreased and interest in and demand for forest products has diminished. The chapter concludes with a discussion of the need for a fundamental rethinking of the future role of community forestry and the governance arrangements which will assist further adaptation.

In the final chapter we draw together the analyses from these chapters to identify key themes, challenges and issues arising for CF in Nepal. A number of major themes emerging from the book are discussed, including issues of power and its distribution and the related challenge of achieving equitable and inclusive decision-making, as well as the challenge for CF in achieving real outcomes to address poverty and marginalization. The evolution of CF in Nepal in response to dynamic and changing local, national and global contexts such as labour migration and the remittance economy is considered, along with the associated new agenda and expanded expectations placed on CF institutions and their implications. While the question of longer term relevance of CF in a dynamic development context is raised, the chapter argues that CF in Nepal will continue to adapt and remain relevant, and that the lessons from the evolution and adaptation of CF in Nepal have global relevance.

Note

1 Gilmour (2016) included 'private ownership (smallholder forestry)' as a fifth point on this spectrum with the strongest rights. However, as private forestry does not really involve 'collective' action, we have chosen not to include it in this spectrum of CBF.

References

Barbier, E.B. and Burgess, J.C. 2001, The economics of tropical deforestation, *Journal of Economic Surveys*, vol. 15, no. 3, pp. 413–433.

Bhattarai, K., Conway, D. and Yousef, M. 2009, Determinants of deforestation in Nepal's Central development region, *Journal of Environmental Management*, vol. 91, no. 2, pp. 471–488.

CBS 2011a, *Nepal Living Standards Survey 2010/11, Statistical Report Vol. 1*, Central Bureau of Statistics, Government of Nepal, Kathmandu, Nepal.

CBS 2011b, *Nepal Living Standards Survey 2010/11, Statistical Report Vol. 2*, Central Bureau of Statistics, Government of Nepal, Kathmandu, Nepal.

CBS 2013, *Statistical Yearbook of Nepal–2013*, Central Bureau of Statistics, Government of Nepal, Kathmandu, Nepal.

CBS 2014, *Statistical Pocketbook of Nepal, 2014*, Central Bureau of Statistics, Government of Nepal, Kathmandu, Nepal.

Dev, O.P. and Adhikari, J. 2007, Community forestry in the Nepal hills: Practices and livelihood impacts, in O. Springate-Baginski and P. Blaikie (eds), *Forests, People and Power: The Political Ecology of Reform in South Asia*, Earthscan, London, UK.

DFRS 2015, *State of Nepal's Forests: Forest Resource Assessment (FRA) Nepal*, Department of Forest Research and Survey, Kathmandu, Nepal, December.

DoF 2017, *CFUG Database Detail*, Department of Forests, January, viewed 8 May 2017, http://dof.gov.np/image/data/Community_Forestry/Detail%20FUG%20All.pdf

FAO 2010, *Global forest resources assessment 2010*, FAO Forestry Research Paper No. 163, Rome, Italy.

FAO 2014, *Global Forest Resource Assessment 2015: Country Report Nepal*, Food and Agriculture Organization of the United Nations, Rome, Italy.

FAO 2016, *State of the World's Forests 2016: Forests and Agriculture: Land-Use Challenges and Opportunities*, Food and Agriculture Organization of the United Nations, Rome, Italy.

Gautam, A., Shivakoti, G. and Webb, E. 2004, A review of forest policies, institutions, and changes in the resource condition in Nepal, *International Forestry Review*, vol. 6, no. 2, pp. 136–148.

Gilmour, D. 2016, *Forty Years of Community-Based Forestry: A Review of Its Extent and Effectiveness*, Food and Agriculture Organization of the United Nations, Rome, Italy.

Kanel, K.R. 2004, *Twenty five years of community forestry: Contribution to millennium development goals*, paper presented at the Fourth National Workshop on Community Forestry, 4–6 August, Kathmandu, Nepal.

Kanel, K.R. 2006, *Current status of community forestry in Nepal*, paper submitted to the Regional Community Forestry Training Centre for Asia and the Pacific, Bangkok, Thailand.

Kanel, K.R. and Kandel, B.R. 2004, Community forestry in Nepal: Achievements and challenges, *Journal of Forest and Livelihood*, vol. 4, no. 1, pp. 55–63.

Keenan, R.J., Reams, G.A., Achard, F., de Freitas, J.V., Grainger, A. and Lindquist, E. 2015, Dynamics of global forest area: Results from the FAO global forest resource assessment

2015, *Forest Ecology and Management*, vol. 352, pp. 9–20. http://dx.doi.org/10.1016/j.
foreco.2015.06.014

LRMP 1986, *Land utilisation report*, Land Resources Mapping Project, His Majesty's Government of Nepal and Government of Canada, Kenting Earth Sciences Limited.

Mahat, T.B.S., Griffin, D.M. and Shepherd, K. 1986, Human impact on some forests of the Middle Hills of Nepal 1. Forestry in the context of the traditional resources of the state, *Mountain Research and Development*, vol. 6, no. 3, pp. 223–232.

Munang, R., Thiaw, I., Rivington, M., Thompson, J., Ganz, D. and Girvetz, E. 2011, *Sustaining Forests: Investing in Our Common Future*, UN Environment Programme (UNEP) Policy series 5–2011, Nairobi, Kenya.

National Planning Commission (NPC) 2014, *An Approach to the Graduation From the Least Developed Country by 2022*, March, National Planning Commission, Government of Nepal, Kathmandu, Nepal.

Nightingale, A. and Sharma, J.R. 2014, Conflict resilience among community forestry user groups: Experiences in Nepal, *Disasters*, vol. 38, no. 3, pp. 517–539.

Ojha, H.R. 2014, Beyond the 'local community': The evolution of multi-scale politics in Nepal's community forestry regimes, *International Forestry Review*, vol. 16, no. 3, pp. 339–353.

Rechlin, M.A., Burch, W.R., Hammett, A.L., Subedi, B., Binayee, S. and Sapkota, I. 2007, Lal Salam and Hario Ban: The effects of the Maoist insurgency on community forestry in Nepal, Forests, Trees and Livelihoods, vol. 17, no. 3, pp. 245–253.

RECOFTC. 2013, *Community forestry in Asia and the pacific: pathway to inclusive development*, Bangkok, Thailand.

Tiwari, K. 2015, A clearing in the trees: Rising demand for firewood due to the quake and fuel crisis could undo Nepal's gains in forest conservation, Letter to *Kathmandu Post*, published 9 December 2015, viewed 27 June 2016, www.forestrynepal.org/article/2184/6403

UN 2015, *Transforming our world: The 2030 agenda for sustainable development*, A/RES/70/1, New York, viewed 8 May 2017, https://sustainabledevelopment.un.org/post2015/transformingourworld/publication

UNDP 2009, *Nepal Human Development Report 2009: State Transformation and Human Development*, United Nations Development Programme, Kathmandu, Nepal.

UNDP 2014, *Nepal Human Development Report 2014: Beyond Geography, Unlocking Human Potential*, National Planning Commission, Government of Nepal, United Nations Development Programme, Kathmandu, Nepal.

UNDP 2016, *Work for human development: Briefing note for countries on the 2015 Human Development report, Nepal*, viewed 8 May 2017, http://hdr.undp.org/sites/all/themes/hdr_theme/country-notes/NPL.pdf

Water and Environment Commission Secretariat (WECS) 2005, *National Water Plan – Nepal*, Water and Energy Commission Secretariat, His Majesty's Government of Nepal, Kathmandu, Nepal.

World Bank 2016, *Personal remittances received (% of GDP)*, viewed 8 May 2017, http://data.worldbank.org/indicator/BX.TRF.PWKR.DT.GD.ZS?locations=NP

2 The history and context of community forestry in Nepal

Robert Fisher, Richard Thwaites and Mohan Poudel

Introduction

In order to understand community forestry in Nepal today, it is useful to understand the context in which it operates and some of the historical background. Nepal has changed greatly in the last few decades. Community forestry has also changed, evolving to adjust to social, economic and political circumstances, as well as responding to international trends and events. A theme of this book is that community forestry continues to evolve and adapt to changing circumstances.

The emergence of community forestry in Nepal

Until the mid-eighteenth century, Nepal consisted of a number of separate small states ruled by local kings (Rajahs). In 1769, following a long series of conquests of the various princely states, Prithvi Narayan Shah, the king of Gorkha, a kingdom in central Nepal, moved his capital to the Kathmandu Valley. The process has been seen as the unification of Nepal. Subsequently, in 1846, the Ranas, members of a powerful lineage, staged a massacre of the royal family and began a period of political control that lasted until 1951 with the restoration of the monarchy under King Tribhuvan. In this period the Prime Ministership of Nepal became hereditary and was held by members of the Rana lineage. The kings of the Shah dynasty remained on the throne merely as figureheads until the restoration of the monarchy.

During the Rana period, in what was essentially a feudal system, areas of land as well as the political control of people living on that land were allocated under various forms of grant to officials and high-ranking aristocrats. In the common form of feudal tenure, these landholders used the land for their own purposes and, in return, depending on which form of grant was involved, sometimes provided a portion of their income earned from tenants as tax revenue to the Rana leaders. The land under this tenure included forests and agricultural land. During the Rana period, land was sometimes cleared by the landholders to open land for agriculture and sometimes to provide timber to the state. The role of providing access to forests and forest products for the tenant was devolved to local officials called talukdars.

In 1951 a revolution took place overturning the Rana rule and restoring the Shah monarchy. In 1960, following some experiments with democracy and a period of considerable political chaos, the king introduced the panchayat system. The panchayat system consisted of several levels, with a national panchayat at the top, district panchayats and village panchayats. It was presented as a traditional or indigenous form of governance based on traditional panchayats or councils (originally consisting of *panch* – five – members), although the traditional panchayats did not exist at the national level, but were actually councils managing the affairs of castes. The system was intended to be a strictly non-party system. Although panchayat members were elected, they were not elected as party members. The system was based on a deeply entrenched ideology, opposed to party politics, and strongly asserted the 'traditional' panchayat concept as the basis of non-political government. In practice, government remained deeply political, based effectively on factions rather than on established parties. Of course parties continued to exist, or were later established, operating covertly.

Village panchayats were local politico-administrative units. They consisted of nine wards and were managed by an elected council headed by a chairperson called the Pradhan Pancha. The panchayats consisted of a number of villages and hamlets and often covered fairly wide areas, particularly in remote hilly regions. Panchayat population was generally in the range of 4,000–6,000 people.

During the period of political chaos that followed the end of the Rana regime, the Nationalization of Private Forest Act was passed in 1957. Although some authors have claimed that this act took control of forests from local people, thus leading to a management vacuum which in turn led to rapidly increasing deforestation, the law was actually concerned with nationalising private forests (that is, those forests held under forms of feudal [private] tenure). The historian Mahesh Chandra Regmi (pers. comm. to Fisher c. 1988) argued that the law was intended as a progressive reform as part of the abolition of the feudal system.

There is no doubt that there was increased deforestation in the aftermath of the revolution of 1950. However, the argument that the nationalization of forests led to the increased deforestation by removing control of forests from local communities is highly questionable. An alternative argument (Gilmour and Fisher 1991) was that the forests deteriorated due to the general chaos and lack of effective government at the time. Fisher et al. (1989) note local accounts of such a crisis at this time. Evidence in support of the theory that deforestation was caused by a crisis following the political change in 1950 includes the fact that 'indigenous' (community-initiated) systems of forest management studied in the late 1980s had often started in the 1960s and 1970s and group members said this happened to protect forests from damage. In other words, they were not old systems, but dynamic responses to changing situations. (For the dating of these indigenous systems, see, for example, Fisher 1989.)

During the 1970s and especially during the 1980s, there was an increasing international response to the issue of deforestation, with the emergence of large numbers of donor-funded projects concerned with forestry issues. Much of the work went into reforestation programs. Amongst the early international projects were the Community Forestry Development Project implemented by UNDP and FAO, the Nepal-Australia Forestry Project, and Swiss- and British-funded projects. To a large extent the emphasis on international funding for forestry was motivated by a narrative about a different crisis, this time environmental rather than political. This narrative presented recent deforestation in Nepal as resulting in landslides, flooding and a general environmental disaster, which ultimately led to massively increased flooding downstream, including in Bangladesh. This crisis narrative was partly inspired by the work of Eckholm (1975), who argued that heavy reliance on fuelwood as domestic fuel was a major factor causing environmental disaster. The documentary film 'The Fragile Mountain' (Nichols 1982) popularized the notion that the Nepali Himalayas were in a massive environmental crisis.

This crisis narrative about the Himalayas, and particularly about Nepal, began to be questioned around the middle of the 1980s. The rethinking of the crisis was synthesized in Ives' and Messerli's (1989) book *The Himalayan Dilemma*. Ives and Messerli examined the grand narrative linking population increase with increased land clearing for agriculture and deforestation due to fuelwood collection, and linked in turn with local flooding and landslides which ultimately led to downstream flooding. They argued that evidence for the links was often weak and that the overall conclusions were flawed. (For a review of some of the research supporting Ives's and Messerli's account, see Gilmour and Fisher 1991). Ives and Messerli referred to the crisis narrative, which they dismissed, as the theory of Himalayan Mountain Degradation.

The early donor projects focused on reforestation, in the form of plantations. The emphasis on reforestation was a response to the crisis narrative of Himalayan degradation. Changing donor priorities over time, such as the later emphasis on poverty reduction and still later on climate change and REDD+, have had a strong influence in changes in forestry policy in Nepal, especially as donor funding provided the lion's share of development funding in the forestry sector, just as it did in development funding in general. (See Chapter 9 for a discussion of aspects of donor-driven forest policy.)

At the same time there was an emerging recognition that involving local communities in forest activities would lead to more effective interventions. This partly reflected an emerging global dialogue on various forms of social (later, community) forestry, exemplified in the book *Forestry for Local Community Development* (FAO 1978). The Nepal government introduced a very early form of community forestry in the form of forestry involving the panchayats. In 1978 regulations were introduced which involved panchayats in two new forms of forest: Panchayat Forests (PF) and Panchayat Protected Forests (PPF). In the case of PF, panchayats were engaged in reforestation activities on areas of land allocated to them. In the case of PPF, panchayats were given responsibility for areas with existing forest. The point about both these programs is that they involved transfer of responsibility for reforestation and protection without providing communities rights to utilize forest products from forests. Technically there was a provision for management and use of forests, but this required approval of an operational plan by the Department of Forests (DoF) and this provision was rarely applied, apart from some experiments.

The PF and PPF programs were essentially about encouraging communities to plant or protect forests on behalf of the DoF. Apart from enthusiasm from some people for environmental objectives, the main incentive for community participation in PF activities came from opportunities for panchayats and some people in them to gain income from nursery and plantation activities. The outcomes of PF in terms of areas of reforestation were, at best, modest. Karmacharya (1987) reports that the total area under PF and PPF in the 29 districts under the Community Forestry Development Project up to 1987 was only 36,376 hectares.

Early community forestry: from PF and PPF to user groups[1]

By 1987 the limitations of PF and PPF were increasingly recognized and the importance of incorporating a focus on livelihoods into community forestry was becoming a concern. This shift in focus led to a major national workshop on community forestry which was held in Kathmandu in 1987. The Decentralization Act 1987 contributed to forming a platform for the review of CF (Krishna Shrestha, pers. comm.). There was a shift in language towards an emphasis on forest user groups as appropriate units for community forestry, as the panchayat was a poor sociological basis for cooperation. However, it became increasingly clear that the political emphasis on the panchayat as the basis of local government complicated the achievement of effective community forestry. We will return to this point.

The shift towards community forestry as a fully-fledged national program became clear with the recommendations of the Master Plan for the Forestry Sector Nepal in 1988 (HMG/N 1988). The outcome of this large-scale exercise of forest assessment and policy planning was a shift towards national policy which promoted handing over forests to communities which had the desire and capacity to manage them.

In the early 1980s there was also an increasing interest in 'traditional' or 'indigenous' systems of forest management. It was recognized that there were indeed cases of traditional forest management protection systems which existed outside the official government-sponsored community forestry program. Reports by Campbell (1978), Molnar (1981) and Messerschmidt (1987) were amongst the first attempts to examine such systems. It is probably true to say that so-called traditional systems were often regarded as interesting remnants and were not seen as being very relevant to contemporary conditions. Studies of traditional systems became increasingly common during the 1980s and into the 1990s, with the features of these systems being seen to be relevant to government-sponsored community forestry.

Research carried out collaboratively by the Nepal-Australia Forestry Project and the International Centre for Integrated Mountain Development (ICIMOD) suggested that indigenous systems were quite common, often effective and, further, that they were often relatively recent developments, rather than simple remnants of systems dating back to the Rana period (see Fisher et al. 1989; Fisher 1989, 1991).[2] The relatively recent origin of these indigenous systems seemed to be largely a response to local people perceiving a vacuum in forest management, due to the limited capacity of the Department of Forests to manage forests effectively, combined with declining forest conditions. This research also confirmed that the user groups for these systems did not match panchayat boundaries, often covering parts of two or more panchayats. The effective user groups did not comprise all residents of panchayats and often came from small sections of adjoining panchayats. This suggests that people with locally and mutually recognized use rights were more likely to form co-operative user groups.

Attempts by the Nepal-Australia Forestry Project in 1986 to prepare an operational plan for Chaap al Danda Forest near Chautara in Sindhu Palchok District highlighted the limitations of assuming that plans should be developed at a panchayat- or multi-panchayat level and the importance of identifying users early in the process. (For discussion of this case, see Gilmour and Fisher 1991.) In large community meetings involving leaders from several panchayats, sections of the forest were allocated to various hamlets for distribution of firewood. However, despite the availability of firewood for collection according to the plan, the scheduled distribution did not work in most cases, largely because it did not reflect the traditional usage patterns; people did not attend because they did not use the particular patches of forest and did not claim use rights. All of this represents evidence for the inappropriateness of panchayat-level approaches to community forestry.

The concept of user group, which officially came into existence through the Decentralization Act 1987, was increasingly accepted as appropriate, but the politically defined requirement to somehow articulate user group forestry with the panchayat system provided challenges until the 1990 revolution. Before turning to the revolution and its implications, it is important to note some important developments of the rapidly evolving community forestry program.

Perhaps one of the most important developments was the wide recognition that community forestry required a shift in the role of forestry field staff away from technical forestry activities and policing forest laws and regulations towards an emphasis on forestry extension and facilitating community development. The shift was not an easy one. Both the work and professional training of forestry field staff had typically been dominated by administration and forest regulation. The policing role frequently led to antagonistic relationships with villagers, especially as bribe taking to allow people access to forest resources was very common and field staff often had serious reputations for corruption.

The consequence of the need to shift the role towards a greater emphasis on community development led to a strong focus on training forest staff. This was often described as 're-orientation training'. Exercises in field staff training became very common from about 1987 onwards, with work by activist and adult educator Kaji Shrestha and his colleagues, including English forester Jane Gronow, being very influential. Shrestha's approach focused on experiential learning and this became widely followed.

It is fair to say that there remained a great deal of scepticism by some government and international project staff about the capacity of some older rangers entrenched in the older corrupt practices to change their approach. In fact, where there was good support from district forest officers, some older rangers found their new role very attractive and rewarding. The challenge was in cases in which range staff did not have support for their community work from their supervisors. One ranger stated that he had been 're-oriented' several times, but was given no authority or support from the District Forest Officer (pers. comm. to Fisher, c. 1987).

A further and related development late in the 1980s and into the 1990s was the increasing interest of many foresters in adopting and supporting community forestry. Part of the appeal of community forestry at this time was the potential it offered for a specific career path. The ever-increasing number of bilateral and multilateral projects provided career opportunities and in many cases this included opportunities for international master's or PhD scholarships. It is surprising how many Nepali foresters have earned international PhD degrees in a variety of fields and how many occupy international positions. This may go some way towards explaining the apparent willingness of many foresters to give up much of their authority over forests in support of community forestry.

Following the acceptance of the Master Plan, there were some experiments with community forestry with user groups. However, it was the revolution of 1990 which largely opened the door for rapid expansion of community forestry. The revolution began with a build-up of discontent about slow national development in early 1990. In only a few months it grew with demonstrations against the panchayat system and the monarchy, culminating with the army firing on a mass demonstration in Kathmandu in late April. A curfew was declared and serious fighting occurred in Patan and elsewhere in the Kathmandu Valley. Although it appeared that the king would be forced to leave Nepal, after several days he announced the end of the panchayat system and agreed to change the political system into democracy under a constitutional monarchy.

While this led to a period of political uncertainty, which in many ways has continued for nearly three decades, it immediately led to the abandonment of the panchayat system and its non-party ideology. Panchayats were replaced by Village Development Committees. While these continued to operate more or less as replacements of the panchayats, the abandonment of the deadening ideological insistence on maintaining the panchayat system created space for more focus on user groups. The number of forest user groups (FUGs) formed increased rapidly.

The situation was subsequently formalized with the Forest Act 1993, which mandated and enabled community forestry in the basic form it has subsequently taken. The major change to community forestry arising from the Forest Act 1993 was a clear mandate for community forestry to operate through forest user groups (subsequently named Community Forest User Groups – CFUGs). Prior to the revolution of 1990 the increasing recognition that user groups were the most appropriate units for community forestry management had been difficult to implement due to the insistence on the panchayat as the lowest level of government and the need to somehow articulate user groups within the panchayat structure.

With the 1993 legislation the emphasis moved clearly to user groups. Under the new law user groups were entitled, subject to certain conditions, to the rights to use and manage community forests. This applied, at least in theory, to all communities with the wish and ability to undertake community forestry. As mandated in the Forest Law 1993, these rights were dependent on making an application to the District Forest Office for the forest to be 'handed over' to the users as members of a CFUG with an approved constitution and use was subject to the approval of an operational plan for forest management. CFUGs were governed by an elected CFUG committee. It is important to emphasize that the approval and handing over of the forest provided only use rights, as agreed under the operational plan, and did not involve transfer of ownership which clearly remained with the state represented by the Department of Forests. This system of tenure remains in place, although the levels of authority granted to the user group has been amended over the years as the forest authorities have attempted to gain increased control over community forests. Indeed, it can be argued that much of the history of community forestry since 1993 has been about attempts by the authorities to restrict the effective power of user groups over community forests against the resistance of some community leaders and activist organizations such as user group federations.

Some programmatic variations of the dominant community forestry model are described in Chapter 6. These include leasehold forestry and collaborative forest management. In these programs the underlying pattern of national ownership with recognized or negotiated use rights persists, but the nature of use rights differs from that in community forestry.

It was intended that the CFUGs would be based on groups of people who were locally recognized as having use rights to forests. These CFUGs often involved residents of several hamlets surrounding a forest. A process of investigation of the appropriate members for a particular CFUG (i.e. those who had locally recognized use rights) was included in operational guidelines.[3]

Despite stress in the guidelines on the need for proper investigation prior to approving CFUG constitutions, it is clear that the CFUG formation process was often rushed in practice, especially as formation of CFUGs and 'handing over' of forests became target driven with the formalization of the community forestry program. Nevertheless the new approach was a great improvement on the older approaches to community forestry.

Following the Forest Act 1993, and particularly with the finalization of the Forest Rules of 1995 which clarified the powers of forest user groups, the number of CFUGs grew rapidly. Figures given by MFSC (2013) show a rapid increase from only 32 CFUGs in 1989 to 2,756 in 1994 and 13,677 in 2005. The rate of growth diminished in the 2000s through the period of the Maoist insurgency and its aftermath, which made it very difficult for government employees to operate in rural areas, yet by 2011 the number of CFUGs had grown to 15,137. Interestingly the rate of growth has jumped again in recent years, with a 24 percent growth between 2011 and 2017 to 19,361 (DoF 2017).

The passing of the 1993 legislation conveniently demarcates a key turning point in the evolution of community forestry in Nepal. Most of the features of the distinctive Nepali approach had emerged and this is the model of community forestry which was increasingly recognized internationally as an example of how to do community forestry. In many countries in Asia, aspects of the Nepali approach were advocated and, in some cases, adopted. In some cases specific terminology has been adopted. For example the term 'community forestry' was adopted in countries as diverse as Bhutan, Thailand and Mongolia and the concept of user group has also been applied (as in Bhutan). Nepali foresters with a background in community forestry have frequently been employed as international advisors on forestry in a number of Asian and African countries.

Community forestry since 1993

The community forestry program has continued to grow since the Forest Act 1993 in terms of the number of CFUGs, the number of people involved and the forest area covered. It has long been a genuine national program. This is not true of most of the countries that have purportedly drawn on the Nepal experience. Nevertheless, there have been issues which have limited the growth and achievements of community forestry in Nepal.

We have already mentioned that many forestry professionals in Nepal in the early days of community forestry were deeply committed to the program, often carving out community forestry-focused careers. However, it is also true that other forestry professionals were not supportive of the community forestry approach and there have been attempts by forestry officials to wind back community forestry, especially in terms of the greater levels of devolution of control to communities that many advocates within the Department of Forest and within civil society were promoting. The process of decentralization remained deeply contested.

One of the most obvious indicators of the contested nature of the process of decentralization in Nepalese community forestry is the history of efforts to impose new regulations on CFUGs after the initial period of growth.

The first visible sign of such efforts to 'put the cat back in the bag' came in 1998 when the GoN (Government of Nepal) passed some amendments to the Forest Act 1993. To summarise:

> The 1998 amendment empowered District Forest Officers to penalize FUG committee members for mismanagement, making them accountable to the District Forest Officer rather than to FUG members. The amendment also required that at least 25 percent of FUG income be spent on forestry development. A directive subsequently issued in 2000 required all FUGs to submit a detailed annual report, including forest inventories to be prepared, for all Operational Plans, both new and under review.
>
> (Fisher 2010, p. 25)

It is important to understand that the shift towards 'upward' accountability to the District Forest Office (DFO) reverses an important principle of decentralization, which is the principle of 'downward' accountability (i.e. the accountability of elected representatives towards their constituents) identified as a key principle of 'democratic decentralization' by Ribot (2002). The second point to note is that the original principle that operational plans be based on simple measures that could be easily and cheaply implemented by CFUGs with the assistance of the District Forest Office staff had become a more onerous and technical requirement, thus shifting management authority away from the CFUGs.

Attempts to further regulate community forestry have continued. Another sign of the reluctance of the forest authorities to allow too large a shift in control towards communities is evident in the reluctance to allow the spread of the community forestry model into the Terai. Community forestry began in the Middle Hills and has largely flourished in the Hills. There has been limited roll-out in the Terai where a modified form of CBFM (community-based forest management) in the form of collaborative forest management has become the program of choice for the Department of Forests. As is described in Chapter 6, collaborative forest management allows the department to maintain far greater management control than community forestry and a substantial share of forest-derived income taken by the government. It may well be that the reason community forestry was allowed to develop in the hills was because the forests, at least at that time, were relatively inaccessible and thus not of great commercial value, unlike the Terai forests (Gilmour and Fisher 1998).

Another indication of the reluctance to allow the full benefits of forest decentralization to flow to communities is the reluctance of the forest authorities to allow CFUGs to harvest and sell timber from community forests in order to generate income. On the other hand, cutting of timber for local use was more easily allowed. Although the legislation does allow CFUGs to harvest timber for sale, gaining administrative approval to allow this is very rare. One (rare) example is the community saw mill, a joint operation between several CFUGs, located at Chaubas east of Kathmandu. This mill has had a chequered history due to changing regulations (Singh 2005; Timsina 2005; Kelly and Aryal 2007).

The various efforts of the forest authorities to regain or maintain control over community forests despite the pressures to decentralize has not occurred without strong resistance. Prior to the 1990 revolution, there were few if any genuine local NGOs (non-government organizations) in Nepal. Since 1990 there has been an intense expansion of civil society representation through NGOs. Many of these NGOs have been active supporters of community forestry and CFUGs.

The most significant development, from the point of view of civil society, has been the development of networks of CFUGs which represent the interests of CFUGs. The most prominent example is the Federation of Community Forestry User Groups, Nepal (FECOFUN) (see Shrestha and Britt 1997; Ojha et al. 2008). Growing out of some local efforts by CFUGs, FECOFUN was registered in 1995 and, with some support from international donor sources, rapidly became a national-level organization, with a complex structure of national and district membership. Its functions include training CFUG representatives in administrative and management skills, as well as advocacy.

This advocacy role has often made forest officials uncomfortable. One of the authors of this chapter (Fisher) attended a training course jointly organized by the Regional Community Forestry Training Center for Asia and the Pacific (RECOFTC), some DoF officials and some NGO representatives. One DoF official (an enthusiast for community forestry) was keen to talk of the supporting 'role' of FECOFUN in implementing community forestry through providing support to the DoF in its implementation role. But he was clearly uncomfortable with the idea that FECOFUN's agenda was advocacy and not determined by government policy. In other words, it saw its role as downward accountability, not upward. Later a separate user group federation, the Nepal Forest Resource User Group, was also formed with support from the DoF. This emerged because FECOFUN was seen to be highly political and critical of the policies and practices of the DoF and some donors, and was lobbying for a greater degree of power transfer to the community level.

The continued process of contestation of control of forests (evident in attempts to control CFUGs) is a dominant theme of post-1993 community forestry. As a collection of papers on decentralization of natural resource management in Asia shows (Wittayapak and Vandergeest 2010), the notion of decentralization as a contested process is not unique to Nepal. The same point is emphasized in a collection on forest reform in India and Nepal (Springate-Baginski and Blaikie 2007).

However the political ecology of community forestry in Nepal does not only relate to contestation over forest control between the state and local communities. It is also quite clear that issues of power are crucial within local communities and CFUGs.

Although democratic governance of CFUGs remains an aspiration for community forestry, the dominance of CFUGs by local elites has been widely reported (see, for example, Shrestha and McManus 2008; Shrestha 2005, 2016). The discussion of elite capture of benefits and control of decision-making revolve around the impacts of domination on marginalized groups in society, broadly identified as women, disadvantaged castes and the poor generally. Putting aside the obvious need to disaggregate the category 'women' into sub-categories, such as high versus low caste women and wealthy versus poor women, the evidence of disadvantage and marginalization is clear.

We will not deal with this point in detail here, as there are examples in later chapters. However, it is important to draw out several important implications.

One outcome of elite domination in some cases has been over-conservative use of forests (Shrestha 2012). This has been partly due to decisions being controlled by local elites (who do not depend extensively on community forests for livelihoods). This occurs despite the clear intent behind community forestry that it addresses poverty and gender bias. There are now regulations that mandate representation by women on CFUG committees and use of a portion of CFUG income for pro-poor programs and activities. As later chapters will show, the outcomes of these regulations are questionable.

The processes by which such elite domination and inequitable livelihood out-comes occur have been discussed. Malla (Malla 2000; Malla, Neupane and Bran-ney 2003) has shown how equal shares of forest products from community forests negatively affect the poor because they have no private resources as alternatives. Wealthier CFUG members can decide on restrictive quotas because they do not depend on community forests. Shrestha (2016) has argued that strict rules of equal sharing result in inequitable (unfair) outcomes, illustrating the process in detail.

As the discussion above shows clearly, issues of power and contested forest resources are central to the realities of community forestry. The history of com-munity forestry in Nepal is largely about the rather paradoxical willingness of many foresters to support the transfer of substantial degrees of control to commu-nities in the years on either side of the passing of the Forest Act 1993, followed by the efforts of many forest authorities to regain control in more recent years. As will be discussed in Chapter 10, the contestation continues in the context of the implementation of the new constitution.

So far, our account of post-1993 community forestry has focused on the forest sector itself. However, the history of community forestry since the 1990 revolution cannot be discussed in isolation from discussion of the political history of Nepal since 1990. Following the brief period of optimism about future development after the revolution, the major participants in the revolution, the Nepali Con-gress Party and the various communist and Maoist factions became increasingly antagonistic towards each other. In the late 1990s there was widespread political unrest in rural areas, with fighting between Maoist and government forces. There were serious atrocities committed by both the Maoists and the national military, and, in many areas, there was no effective government outside district capitals.

By some accounts CFUGs continued to function. Although the situation later deteriorated, the then-Director General of the Department of Forests around 1999 or 2000 told one of the authors (Fisher) that the Maoists had told him that they liked community forestry and would not interfere with it.

In 2001 King Birendra and many members of his family were murdered in the Royal Palace by the Prince. In the aftermath the Maoist uprising expanded and the country entered several years of full-scale civil war. When peace agree-ments gradually developed, the new Maoist government demanded the abolition of democracy and the last king, Gyanendra, was removed in 2008.

Since the breakout of peace, the political situation has been dominated by revolving governments and long-delayed efforts to agree to a new constitution. A new constitution was finally ratified in 2015, although many details remain unresolved at the time of writing and there have been ongoing, including violent, disputes about details of implementation.

This, of course, is a very simplified outline of a very complex political period. The main impact on community forestry of virtually two decades of continuing chaos, of varying levels of intensity, has been that there has been no elected local government for most of the period and very little effective administrative govern-ment at the local level. This has meant that CFUGs have operated as one of the few functioning local institutions and have been affected more by relationships with the forest authorities and other central government agencies, as well as with a variety of NGOs.

Changing context and priorities: concluding remarks

We have already described the shift from an early emphasis on addressing forest loss and degradation towards greater emphasis on meeting livelihood needs. This emphasis was evident in the Forest Act 1993. In the last decade or so there has been a further shift towards discussion of the role of community forestry as a contributor to poverty reduction. However, as is discussed in Chapter 4, there has not been a great deal of achievement in this area.

The limited success in addressing poverty reduction is at least partly related to the tendency of the forestry authorities to emphasize forest conservation over the active (and sustainable) utilization of forests by communities for income generation despite the common perception that forests are under-utilized (discussed in Chapter 9). This has been evident from the conservative nature of most operational plans and especially from the reluctance of the forest administration to approve and support community-based saw mills.

In addition to the impacts of political events, community forestry has evolved in the context of international factors and domestic and international economic change.

Internationally, community forestry has been forced to respond to new agendas such as concerns with climate change, REDD+ and biodiversity priorities. To a considerable extent these new agenda have been driven by international priorities. Community forestry has always been driven by, or at least strongly influenced by, donor priorities, which themselves respond to international policy priorities. Concerns with meeting the challenges of climate change, including the demands of national-level REDD+ policies and pilot projects, as well as contributing to biodiversity targets, have put additional demands on CFUGs. Some of these demands may fit easily with CFUG operations, but others, such as REDD+, present challenges in terms of meeting the original goals of community forestry. It remains to be seen whether local institutions developed with one set of functions can adapt to the wide variety of new demands.

Domestic economic change has contributed to major changes in the labour force as many Nepali men of working age (as many as 4 million) have migrated internationally for employment. The result has been a substantial remittance economy and changes in the practice of agriculture, which involve greater female agricultural labour and apparent decreasing demand for forest products for domestic consumption. These impacts are discussed in more detail in Chapter 9.

Community forestry emerged in the context of a country where substantial numbers of people lived in relatively isolated communities largely dependent on agriculture and natural resources, especially forests, for their livelihoods. The national and international context has changed and the role of community forestry has evolved with the changing context. Part of the intention of later chapters in this book is to explore the changes and to reflect on the possible future directions of community forestry.

Notes

1 Much of the account of the evolution of community forestry in the 1980s and early 1990s in this chapter reflects the experiences and interpretation of the first author of this chapter (Fisher), who worked with the Nepal-Australia Forestry Project between 1987 and 1989 and who continued to be involved for several years. The account is based on personal observation and reflection. For an official view of the history and achievements of community forestry in Nepal, see MFSC (2013).
2 For a somewhat later study of indigenous systems and for an overview, see Tamang, Gill and Thapa (1993). See also Fisher (1994) for an analysis of the relationship between indigenous systems and common property theory.
3 Draft operational guidelines were initially prepared as early as 1988, but revised guidelines were prepared after the 1993 legislation. Guidelines for the CF development program were formally introduced and implemented in 1996. Since then the guidelines have been revised twice (in 2001 and 2009) based on unfolding issues, experiences and learning.

References

DoF 2017, *CFUG Database Summary*, Department of Forests, Kathmandu, January, viewed 23 May 2017, http://dof.gov.np/image/data/Community_Forestry/All%20District%20Summary.pdf

Eckholm, E. 1975, *The Other Energy Crisis: Firewood*, Worldwatch Institute, Washington, DC.

FAO 1978, *Forestry for Local Community Development*, Food and Agricultural Organization of the United Nations, Rome, Italy.

Fisher, R.J. 1989, *Indigenous systems of common property forest management in Nepal*, Working Paper No. 18, Environment and Policy Institute, East-West Center, Honolulu, HI.

Fisher, R.J. 1991, *Studying Indigenous Forest Management Systems in Nepal: Toward a More Systematic Approach*, Environment and Policy Institute, East West Center, Honolulu, HI.

Fisher, R.J. 1994, Indigenous forest management in Nepal: Why common property is not a problem, in M. Allen (ed.), *Anthropology of Nepal: People, Problems and Processes*, Mandala Book Point, Kathmandu, Nepal, pp. 64–81.

Fisher, R. J. 2010, *Devolution or persistence of State control*, in *The Politics of Decentralization: Natural Resource Management in Asia*, in C. Wittayapak and P. Vandergeest (eds) Mekong Press, Chiang Mai, Thailand, pp. 21–35.

Fisher, R.J., Singh, H.B., Pandey, D.R. and Lang, H. 1989, *The Management of Forest Resources in Rural Development: A Case Study of Sindhu Palchok and KabhrePalanchok Districts on Nepal*, International Centre for Integrated Mountain Development (ICIMOD), Kathmandu, Nepal.

Gilmour, D.A. and Fisher, R.J. 1991, *Villagers, Forests and Foresters: The Philosophy, Process and Practice of Community Forestry in Nepal*, Sahayogi Press, Kathmandu, Nepal.

HMG/N 1988, *Main report*, Master Plan for the Forestry Sector Nepal, Ministry of Forest and Soil Conservation, His Majesty's Government of Nepal.

Ives, J.D. and Messerli, B. 1989, *The Himalayan Dilemma: Reconciling Development and Conservation*, The United Nations University and Routledge, London and New York.

Karmacharya, S.C. 1987, Community forestry management: Experiences of the Community Forestry Development Project, *Banko Janakari*, vol. 1, no. 4, pp 30–36.

Kelly, M. and Aryal, P. 2007, Managing the risks of community-based processing: Lessons from two community-based sawmills in Nepal, in R. Obendorf, P. Durst, S. Mahanty, K. Burslem and R. Suzuki (eds), *A Cut for the Poor*, Proceedings of the International Conference on Managing Forests for Poverty Reduction: Capturing opportunities in forest harvesting and wood processing for the benefit of the poor, Ho Chi Minh, Vietnam, 3–6 October 2006, FAO RAP publication number and RECOFTC Report No. 19, FAO and RECOFTC, Bangkok, Thailand.

Malla, Y.B. 2000, Impact of community forestry on rural livelihoods and food security, *Unasylva*, vol. 202, no. 52, pp. 37–45.

Malla, Y.B., Neupane, H.R. and Branney, P.J. (2003). Why aren't poor people benefitting more from community forestry? *Journal of Forest and Livelihood*, vol. 3, no. 1, pp. 78–93.

Messerschmidt, D.A. 1987, Conservation and society in Nepal: Traditional forest management and innovative development, in P.D. Little, M.M. Horowitz and A.E. Nyerges (eds), *Lands at Risk in the Third World: Local Level Perspectives*, Westview Press, Boulder, CO, pp. 373–397.

MFSC 2013, *Persistence and Change: Review of 30 Years of Community Forestry I*, Ministry of Forests and Soil Conservation, Kathmandu, Nepal.

Nichols, S. 1982, *The Fragile Mountain* (Film), Sandra Nichols Productions, Newport, Rhode Island (Sandra Nichols, Director).

Ojha, H., Khanal, D.R., Sharma, N., Sharma, H. and Pathak, B., 2008, Federation of community forestry user groups in Nepal: An innovation in democratic forest governance, in B. Fisher, C. Veer and S. Mahanty (eds), *Poverty Reduction and Forests: Tenure, Market, and Policy Reforms*, Proceedings of an International Conference 3–7 September 2007, Bangkok, RECOFTC and Rights and Resources Initiative, Bangkok and Washington, DC, pp. 1–30 Paper 26 in CD-Rom attached to Proceedings.

Ribot, J.C. 2002, *Democratic Decentralization of Natural Resources: Institutionalizing Popular Participation*, World Resources Institute, Washington, DC.

Shrestha, K.K. 2005, *Collective action and equity in Nepalese community forestry*, PhD Thesis, University of Sydney.

Shrestha, K.K. 2012, Towards environmental equity in Nepalese community forestry, in F.D. Gordon and G.K. Freeland (eds), *International Environmental Justice: Competing Claims and Perspectives*, ILM Publications, Hertfordshire, UK, pp. 97–111.

Shrestha, K.K. 2016, *Dilemmas of Justice: Collective Action & Equity in Nepal's Community Forestry*, Adroit Publishers, New Delhi, India.

Shrestha, K.K. and McManus, P. 2008, The politics of community participation in natural resource management: Lessons from community forestry in Nepal, *Australian Forestry*, vol. 71, no. 2, pp. 135–146.

Shrestha, N.K. and Britt, C. 1997, Crafting community forestry: Networking and federation-building experiences, in M. Victor and J. Bornemeier (eds), *Community Forestry at a Crossroads: Reflections and Future Directions in the Development of Community Forestry*, Proceedings of an international seminar held in Bangkok, Thailand 17–19 July 1997, RECOFTC, Bangkok, Thailand, pp. 133–144.

Singh, H. 2005, Chaubas-Bhumli community Sawmill: Empowering local people, in P.B. Durst, C. Brown, H.D. Tacio and M. Ishikawa (eds), *In Search of Excellence: Exemplary Forest Management in Asia and the Pacific*, Food and Agriculture Organization of the United Nations, Regional Office for Asia and the Pacific and Regional Community Forestry Training Center for Asia and the Pacific, Bangkok, Thailand, pp. 135–144.

Springate-Baginski, O. and Blaikie, P. (eds) 2007, *Forests, People and Power: The Political Ecology of Reform in South Asia*, Earthscan, London and Serling VA.

Tamang, D., Gill, G.J. and Thapa, G.B. (eds) 1993, *Indigenous Management of Natural Resources in Nepal*, HMG Ministry of Agriculture and Winrock International, Kathmandu, Nepal.

Timsina, N.P. 2005, *Supporting livelihoods through employment: The Chaubas-Bhumlu community sawmill, Nepal*, Report prepared for ITTO, Forest Trends, RECOFTC and Rights and Resources.

Wittayapak, C. and P. Vandergeest, P. (eds) 2010, *The politics of decentralization: natural resource management in Asia*, Mekong Press, Chiang Mai, Thailand.

3 Community forestry in Nepal

Analysis of environmental outcomes

Ridish K. Pokharel, Krishna R. Tiwari
and Richard Thwaites

Introduction

Forests are important in a mountainous country such as Nepal for the protection of water catchments and soils, conservation of biodiversity and the maintenance of the environment. Forests in Nepal are also an integral part of the farming system on which the majority of the population is dependent. Whitemand (1980) estimated that up to 50 hectares of forests and grazing is required to maintain one hectare of paddy in high-altitude areas of Nepal. Similarly for the Middle Hills, Wyatt-Smith (1982) proposed that one hectare of farmland requires 3.5 hectares of forest. Over time, with population growth in Nepal, and as a result of government policies under the Rana regime, these important forests have been cleared and converted to agriculture (Soussan, Shrestha and Uprety 1995).

It is difficult to identify the exact area of forest lost in Nepal as there was no forest inventory data before the 1960s. The Ministry of Forests and Soil Conservation and Government of Nepal have carried out forest resource assessments since the 1960s (Acharya and Dangi 2009). The first forest inventory was conducted from 1963 to 1967 by the Forest Resource Survey Office using visual interpretation and mapping of aerial photography from 1953–1958 and 1963–1964 supported by field inventories. A second forest resource assessment was undertaken in 1977–1979 through analysis of aerial photographs combined with ground truthing by helicopter, land surveys and topographic maps. The Department of Forest and Research Survey under the Ministry of Forest and Soil Conservation conducted a third national forest inventory in 1994 using Landsat satellite images, aerial photographs and field measurements.

Over the latter half of the twentieth century, deforestation and forest degradation have been recognized as a problem experienced in many developing countries. Authors such as Kammerbauer and Ardon (1999) and Van Laake and Sanchez-Azofeifa (2004) recognized that human land use decisions have exacerbated deforestation and forest degradation. Between 1964 and 1985, 570,000 hectares of forest was lost in Nepal, coupled with a decline in forest quality as forests were converted into shrubland (Pokharel, Stadtmuller and Pfund 2005). The World Bank (1991) estimated that between 1980 and 1985 deforestation was occurring at an annual rate of 4 percent. The government's efforts in controlling forest loss and degradation failed and could not slow down the rate of deforestation as expected. The National Forest Inventory (DFRS 1999) found a reduction in total area of forest in Nepal from 42.7 percent of land area in 1979 (consisting of 38 percent forest and 4.7 percent shrubland) to 39.6 percent in 1994 (29 percent forest and 10.6 percent shrubland).

With this deforestation and forest degradation, local people in rural areas of Nepal experienced a scarcity of fuelwood for energy, a reduction in the supply of fodder and leaf-litter manure, and a reduction in water quality and supply; these changes in forest condition have also been linked with increasing soil erosion, landslides and downstream flooding (Pokharel, Stadtmuller and Pfund 2005; Chaudhary, Upreti and Rimal 2016; Eckholm 1975). These downstream results of deforestation have been seen by some authors as comprising what Ives and Messerli (1989) referred to as the "theory of Himalayan environmental degradation", linking for example population growth in the hills of Nepal to flooding on the Gangetic Plain. However, Ives and Messerli, as well as a number of scientists at the influential Mohonk Mountain Conference held in 1986, rejected the claims of the disaster theory as grossly overstated (Ives and Messerli 1989; see also Fisher 1990 and Chapter 2 of this book). Despite this debate, the loss and degradation of forests in Nepal clearly presents many challenges, particularly to forest-dependent people.

Forest degradation was seen as a policing as well as a technical problem, so an approach of bureaucratic control seeking to manage and protect forests by applying technical solutions such as plantations and fencing was dominant until the late 1970s. However, Nepal continued to experience loss and degradation of forests, so in seeking to improve management and protection of forests, an alternative approach was gradually implemented which adopted the concept of decentralizing forest management through community-based forestry. From 1978, the government handed over limited forest land to panchayats, a practice that excluded forest users living beyond the administrative boundaries of the panchayats. From about 1988, the emergence of the concept of forest user group management was initiated and applied in pilot projects and, since the passing of the Forest Act 1993, user group community forestry has been formalized in Nepal in managing community forests (see Chapter 2 of this volume for a more detailed explanation of this process).

This chapter explores the contribution of community forestry to environmental outcomes in Nepal, based on the understanding that the initial intention of implementing community forestry was to address environmental problems by reducing deforestation and forest degradation. Over recent decades, a narrative has developed in professional discussions, national news/media and some research around the positive contribution of the community forestry program in solving the environmental problems, particularly reducing the deforestation rate and improving forest condition. In 2015, the Department of Forest Research and Survey produced a State of Nepal's forests report (MoFSC 2015) which (despite methodological differences adopted in different studies) suggests improvements in forest cover over time. In this context, this chapter explores the literature to consider the evidence for the environmental outcomes arising from community forestry.

The following section of this chapter introduces the environmental objectives of Nepal's community forestry, and the third section brings together evidence from the published literature to explore the environmental outcomes of the implementation of the community forestry program in local communities in Nepal.

Environmental objectives of community forestry

Given the continuing trend of deforestation and forest degradation under management by government forest authorities, an alternative concept of community-based forest management has been adopted in many countries with management responsibility passed to local people. Community-based forestry was conceived as passing responsibility for forest protection, restoration, management and utilization to local people (Chhetri and Pandey 1992). In Nepal, an explicit concern with environmental deterioration was expressed in the sixth Five Year Plan (1980–1985), which specified a number of policy measures for conservation and sustainable use of natural resources. This concern was reiterated in the Eighth (1992–1997) and the Ninth (1997–2002) plans as well.

With the failure of state-controlled forest management practice, the Government of Nepal adopted community-based forestry formally in 1978 under the Panchayat Forest and Panchayat Protected Forest Rules and Regulations. The main objective identified for community-based forestry is the restoration of degraded forest land and abating environmental degradation (Adhikari 2005; Gautam 2009), while the implementation of the community forestry program also seeks to enhance forest conservation (Pokharel, Stadtmuller and Pfund 2005; Ojha, Persha and Chhatre 2009) and address environmental degradation by applying a sustainable basis for utilization of forest products (Gautam 2009). In order to achieve the abovementioned objectives, the Ministry of Forest and Soil Conservation set a target of creating Panchayat Forests and Panchayat Protected Forests in their annual plans. The Five Year Plan also set a target for afforestation. For instance, a target of 30,000 hectares was set for afforestation in the ninth plan (Chapagain, Kanel and Regmi 1999). The sixth Five Year Plan had set a target for community forestry program as the protection and improvement of 82,189 hectares of forest land through community forestry (Shrestha and Amatya 2001).

The regulations of Panchayat Forests and Panchayat Protected Forests included the provision for transfer to panchayats of a limited area of degraded government-owned forest (up to 125 ha) and existing natural forest (up to 250 ha). With the establishment of Panchayat Forests and Panchayat Protected Forests, panchayats were to be responsible for planting, maintenance and protection of those forests, with the government providing free seeds and seedlings to the panchayat. This approach was adopted to achieve the objective of increasing forest cover and reducing deforestation. In the early 1980s the government also banned the felling of selected species such as *Shorea robusta, Acacia catechu* and *Michelia champaca* to protect the environment and preserve wildlife and biodiversity (Malla 2001).

In 1980, the Government of Nepal implemented the first national-level community forestry project covering 29 districts in the Middle Hills with the aim of improving environmental conditions (Manandhar 1981). The major activities of the project were afforestation and reforestation, promoting the establishment of forest nurseries and plantations. Free tree seedlings were provided by the government to promote establishment of plantations on community and private lands. In community lands, protection from grazing and theft was required in the early stages, as free gazing in forest and common lands was common practice in rural areas. So, local people were invited to participate in protecting their plantations. However, there was little motivation for people to effectively protect the plantations, as the invitation to participate did not consider their livelihoods. As a result, the success rate of plantations was found to be poor; in many cases planted trees died and targets were not achieved. This was one reason why the policy of forest management through panchayats was not sustained in the long term (Kanel and Dahal 2008). A total area of 35,300 hectares of plantation was established in the Middle Hills and in the High Mountain zones up to 1988 (Gilmour and Fisher 1991).

In many cases, once grazing and illegal harvesting were controlled, species that had been lost reappeared in planted areas (Gilmour and Fisher 1991; Pokharel 2012). As people became aware of the regeneration occurring with control of grazing and illegal harvesting, the focus of management shifted towards natural regeneration. Tree plantation was carried out only in areas where it was badly needed. Most of the community forests in Nepal are natural forests, but manmade forests (plantations) have also been given to Community Forest User Groups. Kanel (2004) reported that some 83 percent of community forest area was covered with forests, 14 percent with shrubs, 3 percent with plantations and less than 1 percent with grass.

Forest policies such as the Master Plan for the Forestry Sector (MoFSC 1988), the Forest Act 1993, and Forest Regulations 1995 were formulated in favour of community forestry with the aim of inviting local people to work together in forest conservation and management. The Master Plan for the Forestry Sector recommended handing over all accessible upland forests to local forest users, while the Forest Act 1993 and Forest Regulations 1995 codified the requirement that Community Forest User Groups (CFUGs) be formed and forest management and product utilization be based on forest operational plans prepared with the technical assistance of the Department of Forests. While this policy was built on the perception of advancing forest loss and degradation, Khadka and Schmidt-Vogt (2008) later identified that the primary objective of community forestry was to meet the basic needs of local people for forest products while managing for multiple purposes. While biodiversity conservation is not specifically mentioned in the Forest Act 1993 or Forest Regulations 1995, there is a requirement for inclusion in the forest operational plan of a 'wild animal conservation provision', and the community forestry program guideline of 2001 introduced a provision for wild animal and biodiversity conservation to be incorporated into forest operational plans (Khadka and Schmidt-Vogt 2008).

Following the shift from a panchayat- to a user group-based approach, many local people took on responsibility for managing forest resources by creation of CFUGs and, in doing so, often adopted protection-oriented management practices by prohibiting the cutting of green trees for a few years, which was found to be an effective approach in improving forest conditions. Shrestha and Amatya (2001) expressed the view that most community forests have been managed for protection and to meet basic needs. The basic forest product needs were fulfilled by felling dead, decaying and dying trees only. In many cases, local people were found to be against cutting green trees in their own forests and formulated the rules accordingly. For instance, Tibrikot Community Forest User Group of Kaski District adopted protection-oriented management practices by putting a ban on harvesting green trees in their forests for some years and allowed only the harvest of dead, decaying, diseased and dying trees when people were in real need (Krishna Kunwar, pers. comm. May 2009). Such practice promoted greenery and generated forest resources along with the improvement of forest conditions, which in turn produced other environmental outcomes.

The national biodiversity strategy and action plan 2014–2020 (MoFSC 2014) has recognized the direct contribution of community forestry to biodiversity conservation and highlights the need to place more attention on managing for biodiversity conservation in CFUG forest operational plans. Following this, a recent Darwin Initiative Project led by the international NGO Birdlife International (Darwin Initiative n.d.) is undertaking a pilot project to 'mainstream' biodiversity by incorporating biodiversity concerns into forest operational plans and CF guidelines, thus placing a renewed emphasis on biodiversity outcomes from CFUG management of forests.

Environmental outcomes of community forestry

A diverse range of ongoing management activities in community forests is expected to not only support sustainable forest management but also to produce various environmental outcomes at the local and national levels through improving forest conditions. For instance, forest protection through patrolling forest areas by local people to control illegal activities in the forest, conservation activities such as tree plantation on open spaces and degraded forest lands, or forest fire management would be expected to enhance environmental outcomes by increasing forest areas, biodiversity and forest resource availability. Similarly, conducting forest inventory and silvicultural operations in community forests and also removing three Ds (diseased, decaying and dying) trees would not only improve the forest quality and productivity but also increase the availability of forest products such as fuelwood, poles and grasses by increasing the number of saplings in the area and tree diameter. This section deals with how community forestry has performed in relation to delivering environmental outcomes and identifies some specific outcomes.

Forest cover

Forest cover is a measure of the area of land with many trees, and such cover plays an important role in reducing runoff (Humbert and Najjar 1992 cited in Meunier 1996). A number of authors (such as Rayamajhi and Pokharel 1998; Pokharel 2005; Pokharel and Suvedi 2007; Pokharel et al. 2015) have identified forest cover as an indicator of forest condition. Community forestry has produced positive results in increasing forest cover. As indicated previously, most community forests have been managed for protection and for providing basic forestry needs with the main objective of allowing the forest to rejuvenate.

Following the narrative of historic forest loss and degradation, the FAO Global Forest Resource Assessment report for Nepal, drawing on a number of forest resource inventories and assessments, shows a steadily decreasing area of forest from 1979 to 2005, and an increasing area of shrubland (which includes degraded forests with no trees and shrub regrowth on abandoned agricultural land) (FAO 2014) (Table 3.1). The results of the third Forest Resource Assessment carried out between 2010 and 2014 presented in the *State of Nepal's forests* report (DFRS 2015a) show a total forest area in Nepal of 5.96 million ha, or 40.36 percent of the total area of the country. Combined with shrublands and other wooded lands (less than 10 percent canopy), the total cover is 6.61 million ha or 44.74 percent of Nepal. The FAO data shows a steady decline in area of forest and increase in area of shrublands, suggesting progressive loss and degradation of the forests. The DFRS data suggests a recent increase in forest cover together with a greatly reduced area of shrublands (suggesting an improvement in forest condition as well as area). This may reflect a real increase in cover, which the report suggests could be a result of community forestry at the local level, or from migration driving the abandonment of agricultural land. While the report notes that both of these explanations are supported by studies at the local level, it advises that further research is required to determine the extent of the contribution of these two factors to vegetation change at a national level. Further, the *State of Nepal's forests* report also offers a third explanation for the apparent increase in forest area: that it is a result of adopting a quite different methodology (DFRS 2015a). Particularly, this assessment has adopted a detailed analysis of satellite imagery supported by field inventory that includes analysis of small patches of forest which were typically not included in earlier studies, thus inflating the total recorded forest area (a number of methodological issues are identified by DFRS [2015b] that mean the recent data cannot be compared to data from the earlier

Table 3.1 Land cover assessment over time ('000 ha)

	1978/79	1985/86	1994	2000	2005	2014
Forest	5593	5504	4268	3900	3636	5962
Shrub	692	706	1560	1753	2235	647
Total	6285	6210	5828	5653	5871	6610
Source	FAO 2014					DFRS 2015a

National Forest Inventory [DFRS 1999]). Table 3.1 shows the time series forest cover assessments from FAO (2014) and DFRS (2015a).

Separate data from the Land Reform Mapping Project also indicates a significant change in land use from 1986 to 2000, with forest area expanding by about 9 percent from 6.2 million ha to 6.8 million ha (FAO 2010). At the same time, agricultural land area was seen to grow by about 20 percent from 3.5 million ha to 4.2 million ha.

Spatial variation in changes in forest cover have also been observed. A 1999 report from the Nepal Department of Forest Research and Survey found widespread loss of forest, with forest cover decreasing in the 12 years from 1978/79 to 1990/91 at an annual rate of 1.3 percent in the Terai and 2.3 percent in the Hills zone (DFRS 1999, cited in Bhuju et al. 2007). A more recent assessment of forest cover by the Department of Forest Research and Survey found that forest area in the Terai (physiographic zone) had decreased at an annual rate of 0.44 percent from 2001 to 2010, and forest in the Siwalik (physiographic zone, also part of Terai ecological zone) had decreased at an annual rate of 0.18 percent from 1995 to 2010 (DFRS 2014). Further evidence for loss of forest cover in the Terai has come from two separate (but more local scale) remote sensing/GIS studies by Kandel (2009) and Pandit (2011). In the Bara District of central Terai, Kandel has found a decrease in forest cover from 36 percent of district area in 1989, to 33.5 percent in 1999, and to 31.7 percent in 2005, a loss of 11.6 percent of initial area in the 16 years from 1989 to 2005. In the Laljhadi forest corridor of the Kanchanpur District of western Terai, Pandit (2011) analyzed Landsat imagery from 1996, 2002 and 2010, describing a fall in forest cover in the 155-square kilometre study area from 63.7 percent in 1996 to 47.7 percent in 2002 and 36.0 percent in 2010, corresponding with an increase in shrubland and degraded forest from 1.4 percent in 1996 to 29.0 percent in 2010. On-ground social research attributed this dramatic loss and degradation of forests to population increase and population encroachment into the forest, driving increased cattle grazing and demand for forest products (Pandit 2011).

A comparison of data sets from the 1989 Master Plan for the Forestry Sector (based on 1985 data) (MoFSC 1988) and from the Nepal Forest Resource Assessment (DRFS 2015a) shows an increase at national scale from 6.2 million ha to 6.6 million ha (Table 3.2), but the patterns become interesting when looking

Table 3.2 Forest cover area by ecological zone ('000 ha)

	DFRS 2015a			MoFSC 1988		
	Forest	Other wooded land	Total	Forest	Other wooded land	Total
Mountain	1,923	553	2,476	1,794	243	2037
Hills	2,254	62	2,316	1,811	404	2215
Terai	1,785	32	1,817	1,913	59	1972
Total	5,962	648	6,610	5,518	706	6224

(Source: DFRS 2015a; MoFSC 1988)

across the ecological zones. In the Middle Hills, total area of forest and shrubland combined has increased by 4.6 percent, but forested land has increased by 24.5 percent, almost entirely a result of a dramatic reduction in area of 'other wooded land' or shrubland. Over the same period of time, total forest and wooded land area in the Terai zone decreased by 7.9 percent, with a decrease in both forest (-6.7 percent) and other wooded lands (-45.8 percent), suggesting that forested land is being replaced by other uses such as agriculture and urban development. These results would support the narrative that community forestry has improved the quality and cover of forests in the Middle Hills, where it has been adopted extensively, but that in the Terai, where community forestry has not been so widely adopted, forests continue to be lost. While different methodologies may make comparison difficult, the different trends seen in the Terai and the Hills are unlikely to be a result only of methodological differences between the two studies, as one would expect that the same methodology was adopted in the Terai as in the Hills in each study.

There are numerous examples of studies that have shown changes in forest cover, based on analysis of local experience under community forestry management, particularly in the Middle Hills. A study conducted using repeat photography by Pokharel, Mahat and Thapa (2011) provides an evocative illustration of a substantial change in forest cover across the landscape over time. The study presents a total of 19 pairs of photos taken in various years and seasons along the way from Kathmandu to Jiri. Baseline photos were taken in various seasons from 1975 to 1985 while more recent photographs were taken from 2005 to 2010. These photos were used to compare changes in forest cover. The study clearly displays that barren/denuded areas and degraded forests have been converted into landscapes covered with lush forests after the implementation of community forestry.

Gautam et al. (2003) assessed the role of community-based management institutions in determining the status of forests in the Upper Roshi Watershed in the Middle Hills Kabre Palanchok District using satellite images from 1976 (Landsat Thematic Mapper satellite), 1989 (Landsat Thematic Mapper satellite) and 2000 (Indian Remote Sensing Satellite). They also used black-and-white aerial photographs from 1978 and 1992 for accuracy and classification estimation to ground truth their analysis. In addition, they also gathered forest-level information on forest type, condition and history of land use from local people and field observation. This study found that 22.5 percent of the 1976 forest area had been lost to other classes by 2000, but had gained 37.4 percent from other classes, resulting in a net 794 ha increase in forest area, or an increase of 5.2 percent of the total watershed area during the study period. They observed an increase in broadleaf forest area from 4,771.4 ha in 1976 to 4,967.1 ha in 1989 and 5,098.4 ha in 2000. Over the same period, agriculture lands had shrunk by about 3 percent between 1979 and 2000. Three explanations were offered for the increase in forest lands: abandonment of agricultural lands on steep slopes; establishment of plantations by the Department of Forests and Community Forest User Groups (1564.5 hectares of plantation were established between 1972 and 1999); and protection under the community forestry program resulting in conversion of grasslands, shrublands and degraded forests into forest. The authors also noted that the period of the study coincided with the inception of the community forestry program across the district, and further noted that the improvement in forest cover was not uniform, with ongoing degradation in government forests in some parts of the watershed while community-managed forests were improving.

Niraula et al. (2013) conducted a study in the Dolakha District (High Mountains ecological zone, but much of which lies within the geographic 'Hills') to understand the changes in forest cover resulting from community forestry over 20 years (1990–2010) using terrestrial photography and satellite remote sensing, coupled with field studies using a hand-held global positioning system (GPS) receiver to identify forest boundaries. The authors compared land cover dividing the district into three clusters (Bhimeshwor, Singati and Thulopatal) consisting of 10 Village Development Committees (VDCs) and 111 Community Forest User Groups. They analyzed change in land cover over about 14,8000 ha of the Bhimeshwor cluster, finding an increase in the area of dense forest from 2,151 ha to 5,643 ha – a 162 percent increase over 20 years. They made similar observations in the Singati and Thulopatal clusters, finding a 97 percent increase in the area of dense forest from 2,690 ha to 5,325 ha in the Singati cluster, and 67.5 percent from 644 ha to 1,078 ha in the Thulopatal cluster. The study also found rates of conversion of non-forest areas to forest, and of sparse forest to dense forest, were considerably higher in community-managed forests than in other forest areas, indicating that improvements in both forest area and density are greater in community-managed forests than in other forests. Their analysis also concluded that community-based forest management has contributed to a decline in slash-and-burn agriculture, reduced the incidence of forest fire, and has resulted in protection and plantation of trees in community-managed forests as well as on private land.

Jackson et al. (1998) examined land use changes in Sindhu Palchok and Kabre Palanchok districts using aerial photographs from 1978 and 1992, as well as ground truthing supplemented by information obtained from local villagers. The samples covered nearly 15 percent of the 400,000 ha area ranging between 600 and 4,000 metres in altitude, including areas covered under the Nepal-Australia Community Forestry Project in the mid-1990s. They identified different trends occurring in different parts of their study area. In the high-altitude upper-slope forest areas, where permanent population is relatively low but there is considerable population pressure from seasonal access to the forests, uncontrolled harvesting of forest products and maintenance of transhumant grazing practices, forest cover is being rapidly denuded, while shrubland and grassland are both increasing (i.e. loss of both forest area and forest quality). In the lower elevation forested slopes, where population is higher and in close proximity to forests, and where collective management had been introduced through community forestry, formerly unproductive shrublands and grasslands had been converted to more productive categories of forested land. They concluded that community forestry activities "are having a beneficial effect on the balance of land use" (Jackson et al. 1998, p. 210).

Similarly, studies based on people's perceptions also indicate that forest cover has changed significantly after the practice of community forestry. People perceive the changes as forest in Nepal is generally in a degraded state when handed over to forest users as community forest. Authors (such as Nurse et al. 2004; Pokharel, Stadtmuller and Pfund 2005; Pokharel and Suvedi 2007; Pandit and Bevilacqua 2011) have reported that greenery in their study areas has improved significantly after the practice of community forestry. Pandit and Bevilacqua (2011) conducted a cross sectional study in the Dhading District of the Middle Hills zone to understand the perception of forest users on the environmental impact of community forestry practices using group interviews and case studies. They used regeneration, forest cover, landslide incidence and soil erosion, water flow in springs and streams, dry season water availability, and wildlife species and abundance as indicators of local environmental conditions. Their findings show strong perceptions from local people of improvements in all environmental indicators (as well as forest product supply indicators), with impressive increases in the number of saplings, poles and trees in only six years under community forests. They argue that such increments have been achieved as a result of conducting regular conservation practices such as grazing control, removal of three Ds trees and tree plantations in degraded areas, as well as carrying out silvicultural activities according to community forest operational guidelines, and conclude that community forestry has been successful in achieving its initial objective of reversing the trend of deteriorating hill environments as well as improving forest product supply to local users.

Forest density

Forest density is a measure of the trees occurring within a particular area, and can be measured in terms such as trees per hectare or basal area per hectare. Evidence shows that forest density has increased following the implementation of the community forestry program. Karna, Gyawali and Karmacharya (2004) conducted a study in seven community forests of five Middle Hill districts (Gorkha, Tanahu, Makwanpur, Kabre Palanchok and Illam) to understand change in forest density from an initial baseline visit in 1993 and follow-up visits 4–6 years later. The study found a mean sapling density of 8,506 per hectare, a significant increment in the density in the community forests from 1993 to 1997. Branney and Yadav (1998) assessed change in forest condition in community forests across 288 research plots in four districts in the Koshi Hills (Middle Hills zone) between 1994 and 1997. They found a significant overall increase over this period of 51 percent in the number of stems per hectare in community forests, as well as a 29 percent increase in basal area in forests in poor starting condition.

Regeneration of forest trees

Natural regeneration of degraded forests provides a pathway to improved forest condition (Tachibana and Adhikari 2009) without the large investment of growing and planting seedlings. Naturally regenerated seedlings are also likely to be quite robust, but regeneration may be threatened by certain management practices such as grazing and fire, or harvesting of timber. Regeneration is commonly measured by recording number of seedlings and saplings of naturally occurring tree species less than two metres tall in a given area (Jackson and Ingles 1998, p. 86), and in the context of regeneration in the Middle Hills of Nepal, Jackson and Ingles (1995) considered regeneration to be good if the number of seedlings are more than 5,000 per ha, and poor if less than 2,000 per ha. Tachibana et al. (2001) argued that identifying a significant improvement in regeneration over time indicates effective management. Based on forest inventory studies in 100 forests from 30 Middle Hill districts in five development regions, Rayamajhi and Pokharel (1998) found an average regeneration in community forests to be 15,685 plants per hectare. Paudyal (2009) carried out a study in Kankali community forest of Chitwan District and found regeneration status of the forest had significantly improved under management as a community forest, with 12,104 plants per hectare. Similarly, a study conducted in 50 community forests of Kaski District by Pokharel (2005) using a nested plot of one square metre found that the average regeneration of the studied community forests as 4,966 saplings per hectare, which he considered to be a sign of good management and improvement of forest condition under community forestry compared to an average measured across the 'hilly area of Nepal' of 1,690 saplings per hectare (FRISP 1999).

Forest condition

Forest condition is a term used by many authors to describe the general condition of forests based on a range of measures such as forest cover, canopy and regeneration, but also including other measures. Improvements in forest condition have been described throughout Nepal after implementation of the community forestry program, contributing to a significant improvement in environmental outcomes such as forest cover, forest area, forest biomass, basal area and forest quality. Tachibana et al. (2001) and Tachibana and Adhikari (2009) examine forest resource conditions under community-based forests using aerial photos taken in 1978 and 1992–1996 and ground truthing in the same number of forest patches. They chose 100 forests from five development regions of Nepal in proportion to the relative importance of forest area in each region compared with total hill area. The study imposed a minimum forest size of 10 ha in order to apply aerial photo analysis with sufficient accuracy. Their findings suggest that forest conditions (crown cover, tree maturity, number of tree species) have improved substantially under community forestry management, as have regeneration rates in the forest. In his abovementioned study of 50 community forests in the Kaski District, Pokharel (2005) also examines forest conditions of the Middle Hills zone using regeneration, crown cover and tree shape as variables of forest conditions and supplemented by information obtained from local people. He identified tree shape as a variable of forest condition because local people refer to tree shape to indicate tree quality (Pokharel 2005). The tree quality is a subjective judgement based on the tree's present condition. The study found good forest conditions for all three indicators measured. The measure of tree quality based on a subjective assessment of tree condition (form, roughness, soundness) found 38 percent of community forests studied with good and 54 percent with medium tree quality. Crown cover was assessed to be high in 74 percent of studied forests and moderate in 24 percent, with an average crown cover across all community forests of 76 percent. Eighty-two percent of assessed forests displayed a high rate of regeneration. Despite the absence of prior data on forest condition, and this study being based on a once only measurement of forest condition, Pokharel (2005) argues that given the degraded nature of many community forests when handed over by the Department of Forests, and the ongoing challenge of over-grazing and cutting of saplings in forests in the hills of Nepal, the presence of good natural regeneration, thick tree cover and good tree shape in community forests indicate that the forest conditions have improved.

Further studies introduced in earlier sections of this chapter have also shown improvement in forest conditions. Branney and Yadav (1998) explored change in forest conditions and management of community forests in the Koshi Hills between 1994 and 1998 describing an overall improvement in community forest conditions. On the basis of their analysis of forest cover change from 1990 to 2010, Niraula et al. (2013) concluded that there were significant improvements in forest condition and that these changes in condition were significantly higher in community-managed forests than other forest areas in all three clusters of Dolakha District. By analyzing the data of tree and sapling density and basal area of five community forests from different districts, Karna, Gyawali and Karmacharya (2004) concluded that there was a significant improvement in forest conditions.

Biodiversity

According to the latest version of the national biodiversity strategy, the "efforts [of community-based forest management] helped abate loss and degradation of forests, and even reversed the trend, particularly in the middle mountains. The improvement in forest condition under community management is believed to have positively contributed to biodiversity" (MoFSC 2014, p. xxiv).

The Government of Nepal's forestry sector policy (MoFSC 2000) gives a high priority to biodiversity conservation, and identified community forests as important areas for improving biodiversity conservation (MoFSC 2014). Evidence shows that by controlling activities such as wildlife hunting, grazing and forest fires, the community forestry program has had positive impacts on biodiversity conservation. Acharya and Oli (2004) investigated shrub and tree diversity using transect walks and discussions with forest users in a community forest in the Parbat District in the Middle Hills of western Nepal. They describe that the land once had only a "few scattered trees" (p. 48), now containing a restored natural ecosystem. Their research found that in 1978 there had been 20 shrub and 17 tree species, but following introduction of protection activities, diversity had expanded to 29 shrub species and 28 tree species in 2003, an increment in the number of tree and shrub species by 60 and 69 percent, respectively. A study conducted by Pandey (2007) showed a higher diversity of tree species on community-managed forests compared to nearby government-managed forests. Nurse et al. (2004) reported that formerly denuded hills are covered with forests and greenery due to community forestry, which has improved plant and wildlife species diversity.

Despite these positive indications for biodiversity arising from community forestry, rural people engaged in subsistence agriculture may not give equal value to all plant species growing in their forest, and giving equal value to all species may not result in the maximum benefits that could be produced from a small number of fast-growing and preferred species (Acharya 2004). This would appear to indicate a contradiction between the objectives of providing forest product benefits to local community and conserving biodiversity. The application of certain silvicultural and harvesting activities in the forest may affect forest composition and structure, resulting in a reduction in biodiversity (Acharya et al. 2006). Some scholars have argued that the community forestry program has reduced plant biodiversity as high-valued timber species are maintained and low-quality timber species removed. Acharya (2004) found that active management of community forests for production of timber threatens the elimination of low-quality timber species and shrub species, supporting the thesis that active management to increase benefits to forest users through extraction of forest products contradicts biodiversity conservation. Acharya (2004) also reported that of 28 tree species in Bharkhore community forest in central Nepal, only 18 were preferred. And in Kali Gandaki community forest, only 15 out of 45 tree species were preferred, and all 33 shrub species identified across the two community forests were categorized as non-preferred species.

In a study of community forestry policies, objectives and practices and their biodiversity implications carried out in the Kathmandu and Nuwakot Districts in the Middle Hills of central Nepal, Khadka and Schmidt-Vogt (2008) found that different forest management activities can have negative and positive impacts on biodiversity; however, common silvicultural practices of removal of species yielding low-quality timber and other unwanted plant species such as thorny bushes and climbers could result in loss of biodiversity in community forests. They argued that given the priority accorded by CFUGs to supply household needs for firewood, fodder, leaf litter, agricultural implements, timber and other non-wood forest products, and lack of awareness of the need for conservation of biodiversity not directly related to their livelihoods, that forest management may have adverse effects on biodiversity. Acharya (2004, Acharya et al. 2006) and Khadka and Schmidt-Vogt (2008) have all argued that in the early stages of community forestry development when the focus is on forest protection and repair of degraded land, community forestry has contributed to biodiversity. But as CFUGs seek to enhance benefits to users through increased forest product supply, silvicultural activities can reduce biodiversity. Acharya et al. (2006) go on, however, to show that active management does not necessarily lead to reduction in species number, finding higher species numbers in an actively managed forest than in a nearby equivalent passively managed forest in the Makawanpur District (Middle Hills), indicating that, if managed appropriately, "harvesting operations may create room for various species other than the dominant species" (Acharya et al. 2006, p. 50).

Availability of forest products and other benefits

Various authors have argued that community forestry management practice has increased the availability of forest products significantly. Research mentioned above by Pandit and Bevilacqua (2011) in the Dhading District (Middle Hills) found an increase in annual availability of fuelwood and fodder in Machhin-dranath community forest of 18 and 16 headloads per hectare respectively, after five years of community forestry practice from 2002 to 2007. Similarly, a smaller increase of eight and three headloads per hectare was found for annual availability of fuelwood and fodder in the Thuloban community forest from 2001 to 2006. Their assessment of change in forest product supply through 22 group interviews across eight community forests also shows a strong perception that all supply indicators (fuelwood, fodder, timber, leaf litter, agricultural implements and other non-timber forest products) had improved under community forestry (Pandit and Bevilacqua 2011). Many other authors (such as Hobley 1996; Branney and Yadav 1998; Springate-Baginski et al. 1999; Gautam, Webb and Eiumnoh 2002, 2004; Kanel, Poudyal and Baral 2005; Pokharel 2005; Nagendra et al. 2008; Tachibana and Adhikari 2009; Oli 2015) have also reported that forest resources availability, availability of diversified forest products and biodiversity have increased due to the community forestry program.

Community forestry has also contributed to improved farm productivity by transfer of nutrients through supply of forest resources such as leaf litter. The harvest of leaf litter from community forests for livestock bedding material and livestock fodder is a major pathway for the nutrient flow from forest to farmland (Pilbeam et al. 2000; Aase, Chaudhary and Veetas 2010; Aase, Chapagain and Tiwari 2013). Leaf litter is also used for making compost by mixing green and dried leaves with animal excreta. Compost is used on farmland, especially in the hills, to replace chemical fertilizer, the frequent use of which can lead to soil degradation, increased soil acidity and reduced productivity. Balla et al. (2014) conducted a study to assess NPK (nitrogen, phosphorus and potassium) content transferred through leaf litter from community forests to farmlands in Hemja village of Kaski District (Middle Hills) and Late and Kunjo villages of Mustang District (High Mountains). They reported on weight of grass and leaf litter collected by households at the time of their research, and calculated the average nutrients transferred from forests to farmlands (28.26 N, 3.21 P and 14.95 K kg per hectare). Interviews with villagers indicated that the amount of leaf litter collected from the forests was higher than five years earlier, resulting in the conclusion that an increase in availability of forest resources under community forestry had enhanced supply of nutrients from forest to farmland.

The three case studies in the High Mountains, Middle Hills and Terai zones reported on by Devkota (Chapter 4 this volume, see Devkota 2010 for full data) have shown a considerable increase in supply of forest products following the introduction of community forests. In what had been degraded government-managed forests experiencing heavy grazing, timber harvesting and intentional fires, supply of a range of products such as timber, fodder and grass, and leaf litter increased markedly under community management. For example, the average supply of leaf litter (in headloads per household per year) across all three case study sites increased by an average of 124 percent from 168 to 376 headloads. The actual increase varied considerably across the three case studies, with a 258 percent increase in the High Mountains Kankali CFUG (Sindhu Palchok District) from 131 to 469 headloads, to 81.5 percent increase in Helejaljale Ka CFUG in the Middle Hills (Kabre Palanchok District) from 184 to 334 headloads, and a 72.6 percent increase in Shreechap Deurali CFUG in Terai (Chitwan District) from 186 to 321 headloads.

Forests are widely recognized as providing a range of direct and indirect benefits for human well-being (Myers 1997; Groot and Meer 2010). For instance, several studies (such as Pokharel, Stadtmuller and Pfund 2005; Pokharel and Suvedi 2007; RRN 2012; Birch et al. 2014; Bhatta et al. 2014) have noted that people perceived the increase in quality and quantity of water for irrigation and water supply, wildlife and habitat as a result of increased forest cover under community forestry management. A study conducted by RRN (2012) in Dharmadevi community forest, Sankhuwasabha District of the High Mountain zone, shows that planting of *utis* tree, bamboos and cardamom in the forest has helped in conserving water and also soil in the area as well as providing income generating opportunities from bamboos and cardamom. Such services provided by improved forests could provide a further opportunity for local communities from environmental outcomes delivered by community forestry management, through the mechanism of Payments for Environmental Services (PES). Such a benefit is potentially available through the PES mechanism as a result of the protection of community forests and provision of different ecosystem services to communities who live downstream. While there is limited experience of application of PES in forests in Nepal, there have been studies of the potential opportunities for forest managers arising from adopting particular management practices.

Bhatta et al. (2014) reviewed 10 PES-type schemes in Nepal, where the PES mechanism is delivering benefits to communities beyond the immediate forest user community, by allowing forest managers to generate income according to the outcomes of their particular management practices. They consider examples of the government-community PES-type mechanism, community-local government PES-type mechanism and community-community PES-type mechanism, mostly related to supply of water (for hydropower, drinking, irrigation, industrial and urban use, as well as for biodiversity and carbon sequestration). Birch et al. (2014) discussed the potential of the Pulchoki Mountain Forest Important Bird and Biodiversity Area (4281 hectares close to Kathmandu in the central Middle Hills area) for delivery of ecosystem services under management as a community forest. Under District Forest Office management, the forest was previously degraded through over-grazing, over-harvesting and uncontrolled fire. Today, one-third of the forest area is managed by 19 community forest groups and has experienced substantial regeneration, with the rest remaining under state management. Based on comparisons of community- and state-managed lands, the ecosystem services were estimated for the forest under community management compared with non-community management, finding that while water provision would not significantly change, soil erosion would increase and water quality would decline, carbon storage would decrease by 64 percent, and there would be changes in harvested wild goods, cultivated goods and recreation value. For those services that were able to be measured in economic terms, the annual net economic value under management as community forest was estimated to be USD800 per hectare per year higher than under the alternative non-community forest management, though this was considered to underestimate the true value because the valuation excluded many services such as water-related services which could not be valued. Others, such as Maharjan (2004), have also discussed the potential of PES to provide economic benefit to community forestry users through delivery of a range of services to external communities, though, as yet, PES arrangements are not prevalent on the ground. Bhatta et al. (2014) provided evidence that PES mechanisms can deliver substantial economic return to local communities, while also providing a means for external stakeholders to influence forest management practices, such as in the case of wildlife conservation or carbon sequestration. Introducing such objectives into community forestry management has the potential to create conflicts with forest product and livelihood objectives of CFUGs (as discussed by Poudel, Rana and Thwaites in Chapter 8 in this volume), requiring careful consultation with forest users and planning.

Conclusions

Community forestry has been found to be an effective approach for forest landscape restoration. There is evidence to show that forest coverage, forest condition, forest density, regeneration and even biodiversity have increased significantly in areas where community forestry has been implemented. Strategies adopted under community forestry focused on protection of forest through reduced grazing pressure and illegal harvesting, managing forest fire and plantations in open space and degraded forest land have led to environmental outcomes such as increased forest area and biodiversity, improved forest conditions, and enhanced forest resource availability and diversity of products. Based on the evidence from a wide range of studies, particularly in the Middle Hills but also experiences in the Terai where community forestry is less prevalent, we can thus conclude that community forestry has contributed significantly to the delivery of positive environmental outcomes, at the local level, across the district landscape level, as well as at the national scale. In recent times, the environmental outcomes of community forestry have been recognized in Nepal, and policy has been developed to further support the delivery of environmental outcomes. Numerous studies have revealed a positive change in forest coverage and forest condition in the Middle Hills as a result of implementation of the community forestry program. Community forest management also includes the management of degraded land through tree plantations, which has facilitated the improvement of land by increasing greenery. While some studies have shown a direct link between community forestry and improved environmental outcomes, there may be alternative explanations for such observations in some situations, such as government-driven plantation projects, abandonment of low-productivity agricultural land or abandonment in cases in which families have migrated away from the district (see Chapter 9 this volume). Biodiversity, too, seems to benefit from the introduction of community forestry, by adoption of more protection-oriented management approaches; however, when user groups turn their focus to active management of forests to better meet the forest product needs of local users, there is potential for biodiversity to be reduced, so careful planning of silvicultural activities will be required if biodiversity conservation is to remain as an objective of community forestry.

References

Aase, H.T., Chapagain, P.S. and Tiwari, P.C. 2013, Innovation as an expression of adaptive capacity to change in Himalayan farming, *Mountain Research and* Development, vol. 33, no. 1, pp. 4–10.

Aase, H.T., Chaudhary, R.P. and Veetas, O.R. 2010, Farming flexibility and food security under climatic uncertainty: Manang, Nepal Himalaya, *Area*, vol. 42, no. 2, pp. 228–238.

Acharya, K.P. 2003, Sustainability of support for community forestry in Nepal, *Forests, Trees and Livelihoods*, vol. 13, no. 3, pp. 247–260.

Acharya, K.P. 2004, Does community forests management support biodiversity conservation? Evidences from two community forests from the mid hills of Nepal, *Journal of Forests and Livelihoods*, vol. 4, no. 1, pp. 44–54.

Acharya, K.P. and Dangi, R.B. 2009, *Case studies on measuring and assessing forest degradation: Forest degradation in Nepal: Review of data and methods*, Forest Resource Assessment Working Paper No. 163, FAO, Rome, Italy.

Acharya, K.P., Goutam, K.R., Acharya, B.K. and Gautam, G. 2006, Participatory assessment of biodiversity conservation in community forestry in Nepal, *Banko Janakari*, vol. 16, no. 1, pp. 46–56.

Acharya, K.P. and Oli, B.N. 2004, Impacts of community forestry in rural households: A case study from Bharkhore community forest, Parbat district, *Banko Janakari*, vol. 14, no. 1, pp. 46–50.

Adhikari, B. 2005, Poverty, property rights and collective actions: Understanding the distributive aspects of common property resource management, *Environment and Development Economics*, vol. 10, pp. 7–31.

Balla, M.K., Tiwari, K.R., Kafle, G., Gautam, S., Thapa, S. and Basnet, B. 2014, Farmers' dependency on forests for nutrient transfer to farmlands in mid-hills and high mountain regions in Nepal (case studies in HemjaKaski, Lete and Kunjo Mustang District), *International Journal of Biodiversity and Conservation*, vol. 16, no. 3, pp. 222–229.

Bhatta, L.D., Helmuth, B.E., Rucevska, I. and Baral, H. 2014, Payment for ecosystem services: Possible instrument for managing ecosystem services in Nepal, *International Journal of Biodiversity Science, Ecosystem Services & Management*, vol. 10, no. 4, pp. 289–299.

Bhuju, U.R., Shakya, P.R., Basnet, T.B. and Shrestha, S. 2007, *Nepal Biodiversity Resource Book: Protected Areas, Ramsar Sites, and World Heritage Sites*, International Centre for Integrated Mountain Development, Kathmandu, Nepal.

Birch, J.C., Thapa, I., Balmford, A., Bradbury, R.B., Brown, C., Butchart, S.H.M., Gurung, H., Hughes, F.M.R., Mulligan, M., Pandeya, B., Peh, K.S-H, Stattersfield, A.J, Walpole, M. and Thomas, D.H.L. 2014, What benefits do community forests provide, and to whom? A rapid assessment of ecosystem services from a Himalayan forest, Nepal, *Ecosystem Services*, vol. 8, pp. 118–127.

Branney, P. and Yadav, K.P. 1998, *Changes in Community Forestry Condition and Management 1994–1998: Analysis of Information From the Forest Resource Assessment Study and Socio-Economic Study in the Koshi Hills, Nepal*, Nepal-UK Community Forestry Project, Kathmandu, Nepal.

Chapagain, D.P., Kanel, K.R. and Regmi, D.C. 1999, *Current Policy and Legal Context of the Forestry Sector With Reference to the Community Forestry Programme in Nepal, a Working Review*, Nepal-UK Community Forestry Project, Kathmandu, Nepal.

Chaudhary, R.P., Upreti, Y. and Rimal, S.K. 2016, Deforestation in Nepal: Causes, consequences and responses, in J.F. Shroder and R. Sivanpillai (eds), *Biological and Environmental Hazards, Risks and Disasters, Hazards and Disasters Series*, Elsevier, Amsterdam, pp. 335–372.

Chhetri, R.B. and Pandey, T.R. 1992, *User Group Forestry in the Far Western Region of Nepal: A Case Study of Baitadi and Achham*, International Center for Mountain Development, Kathmandu, Nepal.

Darwin Initiative n.d., *Mainstreaming biodiversity and ecosystem services into community forestry in Nepal*, viewed 26 June 2017, www.darwininitiative.org.uk/project/22018/

de Groot, R.S. and van der Meer, P.J. 2010, Quantifying and valuing goods and services provided by plantation forests, in J. Bauhus, P.J. van der Meer and M. Kanninen (eds), *Ecosystem Goods and Services From Plantation Forests*, Earthscan, London, pp. 16–42.

Devkota, B.P. 2012, *Socio-economic outcomes of community forestry in Nepal: Lessons from three diverse rural communities*, PhD, Charles Sturt University, viewed 16 May 2017: http://primo.unilinc.edu.au/primo_library/libweb/action/dlDisplay.do?vid=CSU2& docId=dtl_csu40107

DFRS 1999, *Forest Resources of Nepal (1987–1998)*, Publication no. 74, Department of Forest Research and Survey, Ministry of Forests and Soil Conservation and Forest Resource Information System Project, Government of Finland, Kathmandu, Nepal.

DFRS 2014, *Churia Forests of Nepal (2011–2013)*, Forest Resource Nepal Assessment Project/Department of Forest Research and Survey, Kathmandu, Nepal.

DFRS 2015a, *State of Nepal's Forests: Forest Resource Assessment (FRA), Nepal*, December, Department of Forest Research and Survey, Kathmandu, Nepal.

DFRS 2015b, *Middle Mountains Forests of Nepal: Forest Resource Assessment (FRA) Nepal*, Department of Forest Research and Survey, Kathmandu, Nepal.

Eckholm, E.P. 1975, The deterioration of mountain environments, *Science*, vol. 189, no. 4205, pp. 764–779.

FAO 2010, *Land Use Policy and Planning*, Food and Agriculture Organization of the United Nations, June, Pulchowk, Nepal.

FAO 2014, *Global Forest Resource Assessment 2015: Country Report Nepal*, Food and Agriculture Organization of the United Nations, Rome, Italy.

Fisher, R.J. 1990, The Himalayan dilemma: Finding the human face, a review essay on 'The Himalayan dilemma: Reconciling development and conservation', by J. Ives and B. Messerli, *Pacific Viewpoint*, vol. 31, no. 1, pp. 69–76.

FRISP 1999, *Forest Resources of the Hilly Area of Nepal 1994–1998* (Publication no. 73), Ministry of Forests and Soil Conservation, Forest Resource Information System Project, Kathmandu, Nepal.

Gautam, A.P. 2009, Equity and livelihoods in Nepal's community forestry, *International Journal of Social Forestry*, vol. 2, pp. 101–122.

Gautam, A.P., Shivakoti, G. and Webb, E. 2004, A review of forest policies, institutions, and changes in the resource condition in Nepal, *International Forestry Review*, vol. 6, no. 2, pp. 136–148.

Gautam, A.P., Webb, E.L. and Eiumnoh, A. 2002, GIS assessment of land use-land cover changes associated with community forestry implementation in the middle hills of Nepal, *Mountain Research and Development*, vol. 22, no. 1, pp. 63–69.

Gautam, A.P., Webb, E.L., Shivakoti, G.P. and Zoebisch, M.A. 2003, Land use dynamics and landscape change pattern in a mountain watershed in Nepal, *Agriculture, Ecosystems and Environment*, vol. 99, no. 1–3, pp. 83–96.

Gilmour, D.A. and Fisher, R.J. 1991, *Villagers, Forests and Foresters: The Philosophy, Process, and Practice of Community Forestry in Nepal*, Sahayogi Press, Kathmandu, Nepal.

Hobley, M. 1996, *Participatory Forestry: The Process of Change in India and Nepal*, Rural Development Forestry Study Guide 3, Rural Forestry Development Network, Overseas Development Institute, London, UK.

Humbert, J. and Najjar, G. 1992, *Influence de la forêt sur le cycle de l'eau en domaine tempéré. Une analyse de la littérature francophone*, CEREG, Strasbourg, France.

Ives, J.D. and Messerli, B. 1989, *The Himalayan Dilemma: Reconciling Development and Conservation*, Routledge, London and New York.

Jackson, B. and Ingles, A. 1995, *Participatory Techniques for Community Forestry: A Field Manual* (Technical Note 5/95), Nepal-Australia Community Forestry Project, Kathmandu, Nepal.

Jackson, W.J. and Ingles, A.W. 1998, *Participatory techniques for community forestry: a field manual*, IUCN, Gland, Switzerland and Cambridge UK, and World Wide Fund for Nature, Gland, Switzerland.

Jackson, W.J., Tamrakar, R.M., Hunt, S. and Shepherd, K.R. 1998, Land use changes in two middle hill districts of Nepal, *Mountain Research and Development*, vol. 18, no. 3, pp. 193–212.

Kammerbauer, J. and Ardon, C. 1999, Land use dynamics and landscape change pattern in a typical watershed in the hillside region of central Hondura, *Agriculture, Ecosystems & Environment*, vol. 75, pp. 93–100.

Kandel, C. 2009, *Forest cover monitoring in the Bara District (Nepal) with remote sensing and geographic information systems*, Masters dissertation, Universidade Nova de Lisboa, viewed 24 May 2017, https://run.unl.pt/bitstream/10362/2316/1/TGEO0001.pdf

Kanel, K.R. 2004, Twenty five years of community forestry: Contribution to millennium development goals, in K.R. Kanel, P. Mathema, B.R. Kanel, D.R. Niraula and M. Gautam (eds), *Twenty Five Years of Community Forestry: Contributing to Millenium Development Goals: Proceedings of the Fourth National Workshop on Community Forestry*, Community Forest Division, Department of Forests, Kathmandu, pp. 4–18.

Kanel, K.R. and Dahal, G.R. 2008, Community forestry policy and its economic implications: An experience from Nepal, *International Journal of Social Forestry*, vol. 1, no. 1, pp. 50–60.

Kanel, K.R., Poudyal, R.P. and Baral, J.C. 2005, Nepal community forestry 2005, in N. O'Brien, S. Matthews and M. Nurse (eds), *First Regional Community Forestry Forum: Regulatory Frameworks for Community Forestry in Asia*, Proceedings of a Regional Forum, Bangkok, Thailand, 24–25 August, pp. 69–83.

Karna, B.K., Gyawali, S. and Karmacharya, M. 2004, Forest condition change: Evidence from five revisited community forests, in K.R. Kanel, P. Mathema, B.R. Kanel, D.R. Niraula and M. Gautam (eds), *Twenty Five Years of Community Forestry: Contributing to Millenium Development Goals: Proceedings of the Fourth National Workshop on Community Forestry*, Community Forest Division, Department of Forests, Kathmandu, pp. 118–123.

Khadka, S.R. and Schmidt-Vogt, D. 2008, Integrating biodiversity conservation and addressing economic needs: an experience with Nepal's community forestry, *Local Environment*, vol. 13, no. 1, pp. 1–13.

Maharjan, M.R. 2004, Payment of environmental services in community forestry, in K.R. Kanel, P. Mathema, B.R. Kanel, D.R. Niraula and M. Gautam (eds), *Twenty Five Years of Community Forestry: Contributing to Millenium Development Goals: Proceedings of the Fourth National Workshop on Community Forestry*, Community Forest Division, Department of Forests, Kathmandu, pp. 371–377.

Malla, Y.B. 2001, Changing policies and the persistence of patron-client relations in Nepal: Stakeholders' responses to changes in forest policies, *Environmental History*, vol. 6, no. 2, pp. 287–307.

Manandhar, P.K. 1981, *Introduction to Policy, Legislation and Program of Community Forestry Development in Nepal (NEP/80/03)*, Field Document #1a, HMG/UNDP/FAO Community Forestry Development Project, Kathmandu, Nepal.

Meunier, M. 1996, Forest cover and floodwater in small mountain watersheds, *Unasylva*, vol. 47, no. 185, pp. 29–37.

MoFSC (Ministry of Forest and Soil Conservation) 1988, *Master Plan for the Forestry Sector Nepal*, Main Report, Ministry of Forest and Soil Conservation, Kathmandu, Nepal.

MoFSC (Ministry of Forests and Soil Conservation) 2000, *Forestry Sector Policy 2000*, His Majesty's Government of Nepal, Ministry of Forests and Soil Conservation, Kathmandu, Nepal.

MoFSC (Ministry of Forests and Soil Conservation) 2014, *Nepal Biodiversity Strategy and Action Plan 2014–2020*, Ministry of Forests and Soil Conservation, Government of Nepal, Kathmandu, Nepal.

MoFSC (Ministry of Forests and Soil Conservation) 2015, *State of Nepal's Forests*, Ministry of Forests and Soil Conservation, Kathmandu, Nepal.

Myers, N. 1997, The world's forests and their ecosystem services, in G.C. Daily (ed.), *Nature's Services: Societal Dependence on Natural Ecosystems*, Island Press, Washington, DC, pp. 215–235.

Nagendra, H., Pareeth, S., Sharma, B., Schweik, C.M. and Adhikari, K.R. 2008, Forest fragmentation and regrowth in an institutional mosaic of community, government and private ownership in Nepal, *Landscape Ecology*, vol. 23, no. 1, pp. 41–54.

Niraula, R.R., Gilani, H., Pokharel, B.K. and Qamer, F.M. 2013, Measuring impacts of community forestry program through repeat photography and satellite remote sensing in the Dolakha district of Nepal, *Journal of Environmental Management*, vol. 126, pp. 20–29.

Nurse, M., Tembe, H., Paudel, D. and Dahal, U. 2004, From passive management to health and wealth creation from Nepal's community forestry, in K.R. Kanel, P. Mathema, B.R. Kanel, D.R. Niraula and M. Gautam (eds), *Twenty Five Years of Community Forestry: Contributing to Millenium Development Goals: Proceedings of the Fourth National Workshop on Community Forestry*, Community Forestry Division, Department of Forests, Kathmandu, Nepal, pp. 127–135.

Ojha, H., Persha, L. and Chhatre, A. 2009, *Community forestry in Nepal: A policy innovation for local livelihoods*, Discussion Paper 0913, International Food Policy Research Institute, Washington, DC.

Oli, B.N. 2015, *Evaluating community forestry processes and outcomes*, PhD, University of Copenhagen.

Pandey, S.S. 2007, *Tree species diversity in existing community based forest management systems in central mid-hills of Nepal*, MSc, Swedish Biodiversity Centre, Upsala University, viewed 3 October 2017, http://www.slu.se/globalassets/ew/org/centrb/cbm/dokument/publikationer-cbm/masteruppsatser/shiva-shankar-pandeythesis.pdf

Pandit, R. and Bevilacqua, E. 2011, Forest users and environmental impacts of community forestry in the hills of Nepal, *Forest Policy and Economics*, vol. 13, pp. 345–352.

Pandit, S. 2011, *Forest cover and land use changes: a study of Laljhadi Forest (Corridor)*, Far Western Development Region, Nepal, MEnvSc, Tribhuvan University, viewed 3 October 2017, http://www.forestrynepal.org/images/thesis/MSc_SantaPandit.pdf

Paudyal, B.K. 2009, Regeneration, plant diversity and growth of hill *Sal* in community forests: A case study from Kankali community forest, Chitwan, in M. Balla, R.M. Bajracharya and B.K. Sitaula (eds), *Natural Resources Management: Reviews and Research in the Himalayan Watersheds*, Institute of Forestry, Pokhara, Nepal, pp. 9–17.

Pilbeam, C.J., Tripathi, B.P., Sherchan, D.P., Gregory, P.J. and Gaunt, J. 2000, Nitrogen balances for households in the mid-hills of Nepal, *Agriculture, Ecosystem and Environment*, vol. 79, no. 1, pp. 61–72.

Pokharel, B., Mahat, A. and Thapa, S. 2011, *Impact of Community Forestry in Nepal Kathmandu to Jiri: A Photo Journey*, revised edn, Nepal Swiss Community Forestry Project, SDC Inter-Cooperation Nepal, Kathmandu, Nepal.

Pokharel, B.K., Stadtmuller, T. and Pfund, J.L. 2005, *From Degradation to Restoration: An Assessment of the Enabling Conditions for Community Forestry in Nepal*, Intercooperation, Swiss Foundation for Development and International Cooperation, viewed 26 April 2017, http://nepalpolicynet.com/images/documents/forest/research/An%20Assessment%20of%20the%20enabling%20conditions%20for%20CF%20in%20Nepal.pdf

Pokharel, R. 2012, Factors influencing the management regime of Nepal's community forestry, *Forest Policy and Economics*, vol. 17, pp. 13–17.

Pokharel, R.K. 2005, Assessing community forests' conditions using variables recommended by local people: A case of Kaski district Nepal, *Banko Janakari*, vol. 15, no. 1, pp. 40–48.

Pokharel, R.K. 2009, Pro-poor programs financed through Nepal's community forestry funds: Does income matter? *Mountain Research and Development*, vol. 29, no. 1, pp. 67–74.

Pokharel, R.K., Neupane, P.R., Tiwari, K.R. and Köhl, M. 2015, Assessing the sustainability in community based forestry: A case from Nepal, *Forest Policy and Economics*, vol. 58, pp. 75–84.

Pokharel, R.K. and Suvedi, M. 2007, Indicators for measuring the success of Nepal's community forestry program: A local perspective, *Human Ecology Review*, vol. 14, no. 1, pp. 68–75.

Rayamajhi, S. and Pokharel, R. 1998, *From deforestation to reforestation: Common property forest management in the hill region of Nepal*, Report submitted to the Environment and Publication Technology Division, International Food Policy Research Institute, Washington, DC.

RRN 2012, *RRN's 20 Years of Experience in Community Based Natural Resource Management*, Rural Reconstruction Nepal, Kathmandu, Nepal.

Shrestha, K. and Amatya, D. 2001, Protection versus active management of community forests, *Community Forestry in Nepal: Proceedings of the Workshop on Community Based*

Forest Resource Management, 20–22 November 2000, pp. 3–17, Joint Technical Review Committee, Kathmandu.

Soussan, J., Shrestha, B.K. and Uprety, L.P. 1995, *The Social Dynamics of Deforestation: A Case Study From Nepal*, Parthenon Publishing Group, New York.

Springate-Baginski, O., Soussan, J.G., Dev, O.P., Yadav, N.P. and Kiff, E. 1999, *Community forestry in Nepal: Impacts on common property resource management*, Working Paper No. 3, Environmental and Development Series, University of Leeds, Leeds, UK.

Tachibana, T. and Adhikari, S. 2009, Does community based management improve natural resource condition? Evidence from the forests in Nepal, *Land Economics*, vol. 85, no. 1, pp. 107–137.

Tachibana, T., Upadhyaya, H.K., Pokharel, R., Rayamajhi, S. and Otsuka, K. 2001, Common property forest management in the hill region of Nepal, in K. Otsuka and F. Place (eds), *Land Tenure and Natural Resource Management: A Comparative Stud y of Agrarian Communities in Asia and Africa*, The Johns Hopkins University Press, Baltimore, MD, pp. 273–314.

Van Laake, P.E. and Sanchez-Azofeifa, G.A. 2004, Focus on deforestation: Zooming in on hot spots in highly fragmented ecosystems in Costa Rica, *Agriculture, Ecosystems & Environment*, vol. 102, pp. 3–15.

Whitemand, P.T.S. 1980, *Agronomy research in the hills areas of Nepal*, Terminal Report, FAO, Rome, Italy.

World Bank 1991, *World development report 1991*, Oxford University Press, New York.

Wyatt-Smith, J. 1982, *The agriculture system in the hills of Nepal: The ratio of agriculture to forest land and the problem of animal feed*, APROSC Occasional Paper No. 1, APROSC, Kathmandu, Nepal.

Yadav, K.P. and Branney, P. 1999, Measuring forest and user group changes in community forestry: Results from the Koshi hills, *Banko Janakari*, vol. 9, no. 1, pp. 20–25.

4 Community forestry, rural livelihoods and poverty reduction in Nepal

Binod Devkota, Richard Thwaites and Digby Race

Introduction

In the formative years of community forestry (CF), most community forests were managed for plantation establishment and forest protection, rather than for forest products to support the livelihoods of rural communities. However, the focus soon turned to conserving not only timber species, but also a wider range of species that provide firewood, fodder and other non-timber forest products (NTFPs), which are especially valuable for subsistence livelihoods of local rural people. In this context, access to forest products for self-consumption was of primary concern. Later, the program developed more explicit concerns with income generation and poverty reduction, partly as a response to international development priorities and partly as a response to changing concerns within Nepal. This chapter explores the extent to which community forestry has contributed to rural livelihoods, income generation and poverty reduction, arguing that the economic benefits have generally been restricted and especially that distribution has been inequitable and biased towards elites within communities. While CF can deliver diverse socio-economic outcomes that may contribute to rural livelihood improvements and poverty reduction, this chapter examines the tangible benefits of community forests such as access and use of forest products and the mobilization of funds from community forestry in supporting rural livelihoods and poverty reduction. The chapter explores issues raised in Chapter 2 regarding ways in which institutional governance, often associated with social and cultural structures around gender and caste, influences distribution of benefits within and between communities.

An overview of community forestry in Nepal as a forest management framework

Forests and trees play an important role in the reduction of rural poverty in many developing countries (FAO 2010). However, many factors may influence the contribution of forest resources to rural livelihoods, including local socio-economic conditions and forest management regimes, opportunities for enterprise development and access to competitive markets for forest products (Adhikari, Falco and Lovett 2004).

It has been widely established that past and present government-managed forestry programs have not been successful in stopping deforestation, and that those programs have not always been successful in supporting the livelihoods of rural communities (Adhikari, Falco and Lovett 2004). The consensus is that a government-centred forest management approach cannot achieve sustainable forest management and poverty reduction (Acharya and Acharya 2007; Malla 2009).

One alternative to government-managed forestry programs is the community-based forest management (CBFM) approach (Malla 2009). CBFM policies have emerged in response to local-level failure of government-managed forestry, and continued deforestation and forest degradation in many developing countries (Devkota 2005; World Bank 2008), with CBFM models typically adopting more participatory decision-making and delegating more functions from central forestry administrations to local governments and communities (Nagendra and Gokhale 2008), such that sustainable economic development can improve the livelihoods of rural poor people. One CBFM approach is the CF model adopted in Nepal. CF provides a framework for rural poor people who use forest resources to exercise control and management over those forest resources as an integral part of their farming system and to meet their domestic requirements.

The shift of control and management of forest resources to local communities to meet their livelihood needs has helped deliver more effective governance of forest resources (see Chapter 2), and as Kanel (2004) has argued, CF can deliver various benefits that include improvement in the biophysical conditions of local forests (i.e. in terms of both quality and quantity), empowerment of local communities to manage forests sustainably and development of local forest-based enterprises and industries.

However, CF in Nepal is not always easily implemented. Some people (e.g. Agarwal 2009; Bhattarai, Jha and Chapagain 2009) argue that policymakers and stakeholders in CF should attend to the removal of barriers, such as:

- inequities regarding the social exclusion of women, poor and disadvantaged groups, and ethnic minorities;
- conflict mismanagement; and
- traditional and centralized government bureaucracy, miscommunication, lack of market access that hinder the flow of benefits to rural communities.

CF in Nepal is considered to be a global innovation in participatory environmental governance in efforts to achieve the twin goals of forest conservation and poverty reduction (Kanel and Dahal 2008). Realizing the need to involve people in preventing deforestation, the government changed its forest policies and strategies, and started handing over forests near rural settlements in the Middle Hills region to local people for protection and use (GoN 1989). In 1978, CF was adopted as a new strategy with initial emphasis on participation of local community user groups in reforestation of degraded land. It has now become a viable option for forest regeneration and protection while providing forest products to meet local needs (GoN 2015; Nagendra and Gokhale 2008).

By the late-1980s, CF had been modified to incorporate rural development concerns into participatory forest management. Participatory forest management involves the transfer of control of local forests to Community Forest User Groups (CFUGs) that hold locally recognized forest use rights. Subject to conditions under the authority of the Department of Forests, the Forest Act 1993 and Forest Rules 1995 provided CFUGs with legal rights to all forest products from their forest, and responsibility to protect and manage the forest (but did not give the rights to sell the land, build houses or cultivate the area) (Kanel and Dahal 2008).

Formal institutions, such as those which formulate government forest policy and legislation, and local CFUGs have key roles to play in relation to the sustainable management and development of forest resources as well as ensuring the flow of goods and services to local rural communities. Some earlier studies indicated that after government-managed national forests were placed in the care of rural communities, the physical condition of those forests (degraded at the time of handover) had generally improved (see Chapter 3), though others have argued that the quantity of forest products which are currently being supplied from the majority of community forests in Nepal is well below the CFUGs' overall demand for these products (Bhattarai, Jha and Chapagain 2009). This raises a question regarding the willingness or ability of CF institutions to deliver livelihood benefits to rural communities, and thus questions the broader socio-economic outcomes of CF.

It is important to note that use of forest products for household consumption (e.g. subsistence use) and forest products being used as the basis for poverty reduction (e.g. tangible commercial benefit) are not the same thing. Poverty reduction involves improving the economic well-being specifically of the poor members of a community, not just a general increase in income provided to the community as a whole.

In an overview of community forestry in Asia, Fisher (2014, p. 17) concludes that 'overall the cash returns (i.e. income) from community forestry have generally been modest'. In addition to this, he stresses that for poverty reduction the poor have to be specifically targeted. Much of the literature on the pro-poor benefits of community forestry in Nepal is not very clear on the extent to which benefits and income are distributed to the poor. On the other hand a number of studies (e.g. Malla 2000; Malla et al. 2003) find clearly that the poor often benefit less than wealthier people. Malla actually argues that the poor may become worse off as a result of community forestry than they were before. This could happen, for example, when a price is imposed on firewood collection that poorer people may not be able to afford.

This chapter reports on the cases of three CFUGs in terms of livelihood benefits and poverty reduction. The findings are consistent with, and thus support, a strong trend in the literature which suggests that livelihood benefits and especially poverty reduction arising from CF in Nepal have been, at best, modest.

Understanding the links between community forestry, rural livelihoods and poverty

In general, a livelihood encompasses people, their abilities and their means of living, including their capacity to source food, shelter, clothes, income and other desirable assets (based on Chambers and Conway 1992 and DFID 2001). Poor people in rural areas often lack the assets to sustain healthy livelihoods. Poor health, lack of clean water and sanitation, limited infrastructure and remoteness from government services are other factors that prevent them from sustaining or rebuilding their livelihoods.

Rural livelihoods in developing countries are directly linked with forest resources. Research conducted on non-farm employment and income in Nepal has indicated increasing interest in the contribution of forests to local rural employment and income (Maharjan and Khatri-Chhetri 2006; Acharya and Acharya 2007). These contributions are important for rural livelihoods, and sustainable forest management ensures that forests continue to make those contributions. Forest resources constitute a safety net for poor rural people in times of hardship or crisis (Kanel 2004). In the absence of state welfare, poor people often rely on nearby forests to provide a means to survive.

Many farming households in Nepal cannot grow enough crops to be self-sufficient all year round. The importance of income from selling forest products is usually determined by time available, and is seldom a large proportion of a household's total income. However, Nagendra and Gokhale (2008) argue that income from forest products may be important, as it often fills seasonal or other gaps in cash flow, and helps people cope with particular expenses or respond to unusual opportunities. The time spent by households on activities in the forest may depend on various factors, including the time of year which may affect the availability of raw materials, the need for extra cash at particular times in the annual cycle (e.g. purchasing of seeds, or hiring of labour), and seasonal fluctuations in demand for or availability of labour for harvesting and processing forest resources.

Bhattarai (2016) and GoN (2015) indicated that rural people, especially poor people, who do not have alternatives for meeting their daily needs, regularly collect forest products for use at home and for generating a small income to support their livelihoods (Table 4.1). In this context, forest resources are often critical elements in farming systems (Nagendra and Gokhale 2008). For poor people, forests provide a way to maintain soil fertility, without resorting to expensive fertilizers. Household income can be augmented by harvesting, processing and selling forest products, such as fuelwood, timber, bamboo, fodder, leaves, flowers, fruits, medicinal plants, honey, tools and meat. In fact, forest resources are attractive for poor people as they are easily accessible, harvesting requires little capital or technical skills, and the produce can usually be processed at home. The evolution of policy instruments worldwide has increased the potential for previously 'indirect benefits' (e.g. protection of land, forest and water resources) to become tradeable services, such as with the advent of various 'payments for environmental services' (PES). Although most PES programs are still experimental or pilots,

Table 4.1 Benefits from community forest resources for poor people

Benefits from forest resources	Examples
Subsistence goods	Wood for building, fuelwood, fodder, bedding materials, medicinal and aromatic plants, honey, mushrooms, leaves, fruits, meat, and so on
Goods for sale	All of the above goods and by-products, including arts and crafts items such as baskets, furniture
Income from employment	Both in the formal and informal sectors
Indirect benefits	Enhancement of land, water, clean air, religious and spiritual sites, cultural heritage, biodiversity conservation, health improvements, environmental goods and services

(Source: Adapted from Bhattarai 2016 and GoN 2015)

they illustrate the policy intent to find ways for local communities to be rewarded for producing a range of environmental services needed by wider society.

Agriculture, as the mainstay of the rural economy in Nepal, provides a livelihood for more than 75 percent of the population (MFSC 2008). More than 83 percent of Nepal's population live in rural areas, and depend on subsistence farming; and about 81 percent of the workforce are employed in agricultural activities, which account for 40 percent of the gross domestic product (CBS 2012 (though see Chapter 9 for discussion of changes to this situation related to labour migration and the remittance economy). An insecure food supply and poor nutrition are major challenges. Most rural households have little or no access to basic services such as primary health care, clean drinking water and sanitation services or education, while the occupants of those households are often illiterate and are landless or have very small landholdings. Nepalese agriculture is characterized by small and fragmented subsistence farming with average landholding less than 0.2 hectares (CBS 2007). The immediate concern is how to increase crop yields, income and food security, as well as infrastructure development to improve livelihood security.

We can see that forests are an integral part of the rural economy and daily lives of rural people in Nepal. They provide employment, value regarding the processing and marketing of forest products, energy, trade and investment back into the forest sector. The majority of rural people use forest products such as fuelwood, fodder, timber, food, medicinal herbs, ritual materials, cattle bedding materials and compost for agricultural fields. Forests are also the main sources of construction materials for housing and agricultural implements. The following section provides a case study of linkages between community forestry and rural livelihoods in Nepal.

Case studies

This chapter reports on findings from detailed PhD research undertaken by Devkota (2012) to explore the livelihood outcomes for local communities arising from CF. Socio-economic characteristics of CFUGs are important factors influencing the implementation of CF and generation of outcomes of the program (Adhikari 2003). The performance of CFUGs in generating socio-economic outcomes from CF also depends on previous history of forest management and the implementation experience of the groups involved (Malla 2009). Recognizing the complexity of how these diverse factors might influence socio-economic outcomes, this research collected data from three CFUGs on their experiences of implementing CF. These CFUGs were purposely selected from three very different districts where CF has been implemented over more than 15 years to cover a broad range of socio-economic characteristics (i.e. population, size, wealth, CFUG membership). The case study districts were Chitwan – an Inner Terai district; Kabre Palanchok – a Middle Hills district; and Sindhu Palchok – a High Mountains district (see Figure 1.2). In each district, one CFUG was selected based on the following criteria:

- have been implementing a CF program for at least 10 years, such that socio-economic outcomes are apparent and can be assessed;
- have completed at least one revision of their forest operational plan (FOP);
- have a high level of dependence on CF and the resulting products;
- have received external (government/NGO) investment to support CF development;
- were identified by the District Forest Office as relatively successful in implementing CF, to enable exploration of what is possible, and the issues faced implementing different strategies and practices;
- were accessible and secure to enable data collection.

Primary data, mostly qualitative in nature, were collected during 2009 and 2011, involving in-depth interviews with 71 key informants, surveys of 138 CFUG member households and six focus-group discussions with poor, women and Dalit members of the CFUGs. This was complemented with analysis of secondary data from journals, books, office records, audit reports, constitutions and forest operational plans of the three selected forest user groups. Additional data for this chapter were collected in 2016.

Social structure plays an important role in shaping local institutions, and rural villages in Nepal typically retain traditional hierarchical structures based on caste, ethnicity and gender. Elites are denoted as the people with substantial advantage in decision-making processes due to their wealth, social/caste and/or hereditary status. Women, as in most other parts of Nepal, hold a lower social position when compared to that of men. Because Nepal is a largely patriarchal society, there are currently fewer women involved in public decision-making processes than men, even though in recent years policymakers have encouraged more participation by women. However, this situation varies from one CFUG to another, because each community differs in terms of wealth, ethnic composition and women's participation. A summary of socio-economic characteristics and forest statistics of the case study CFUGs is presented in Table 4.2.

Table 4.2 Socio-economic and forest attributes of case study CFUGs

Characteristics	Kamkali CFUG (KCFUG)	Hilejaljale 'Ka' CFUG (HJCFUG)	Shreechhap Deurali CFUG (SDCFUG)
District	Chitwan	Kabre Palanchok	Sindhu Palchok
Year of forest handover	1995	1998	1998
Forest area (hectares)	761	118	81
Elevation (m)	300–900	1810–2080	1200–1600
*Heterogeneity of community in terms of caste and ethnicity	Most heterogeneous	Least heterogeneous	Medium
	UC – (Brahmin and Chhetri)	UC – (Brahmin and Chhetri)	UC – (Brahmin and Chhetri)
	MC – (Tamang, Gurung, Magar, Newar, Tharu, Kumal, Majhi, Darai, Dhami, Chepang)	MC – (Bhujel)	MC – (Newar and Tamang)
	LC – (Damai, Kami and Sarki)	LC – (Damai and Kami)	LC – Kami
Total households (HH)	1832	250	247
HH categories	poor: 735, medium: 835, rich: 262	poor: 90, medium: 75, rich: 85	poor: 121, medium: 78, rich: 48
Total population	12172 (male: 6070, female: 6102)	1550 (male: 773; female: 777)	1490 (male: 748; female: 742)
Forest area per HH (ha)	0.41	0.47	0.33
Average HH size	6.6	6.2	6.4
CFUG EC membership in 2012	17 (male: 11; female: 6) (UC: 12; MC: 4; LC: 1)	11 (male: 7; female: 4) (UC: 10; MC: 0; LC: 1)	11 (male: 7; female: 4) (UC: 4; MC: 6; LC: 1)
Literacy (%)	72 (male: 79; female: 65)	70 (male: 74; female: 65)	69 (male: 72; female: 65)
Main occupation of users (%)	agriculture: 78%; service: 5%; small business: 3%; labour: 14%	agriculture: 80%; service: 4%; small business: 5%; labour: 11%	agriculture: 70%; service: 7%; small business: 4%; labour: 19%
Forest statistics (based on forest inventory data prior to handover and 2012 forest inventory)			
Area (ha)	760	118	124
Canopy cover (%) (before/after CF)	40/60	60/70	60/75
Average (tree/ha) (before/after CF)	No data/1700	1800/3400	No data/600
Average seedlings (/ha) (after CF)	9,984	3,700	6,300
Productivity after CF (average annual increment), m³/ha	38	10	11
Forest type	Natural *Shorea robusta* and *Dalbergia sissoo – Acacia catechu* plantation	Pine plantation forest with natural broad-leaved species	Pine plantation forest with natural broad-leaved species
Changes in number of plant and wildlife species after CF	Increased	Increased	Increased

Note: EC: executive committee; UC: Upper caste; MC: Middle caste; LC: Lower caste; HH: households.
(Source: compilation of office records and FOPs of the CFUGs 2012)

Table 4.3 Criteria developed by CFUGs for wealth ranking

Wealth categories	Landholding size and other criteria
Poor	less than 6-months food supply from own land; average annual per capita income is less than NRs 25,000
Medium	up to 9-months food supply from own production; or average annual per capita income is less than NRs 35,000
Rich	food supply for 12 months or average annual per capita income is more than NRs 35,000

(Source: fieldwork 2009 [participatory ranking exercise])

A participatory wealth ranking exercise was conducted at each case study site to identify the economic status of households of participants in the various CFUGs based on their own criteria (Table 4.3).

Community access to forest products

Recognizing that these case study communities are heterogeneous, and that with elite status comes some element of power and authority, CF policy seeks to ensure that adequate resources flow to those who need it most. Any improvement regarding equity implies an improvement in the capabilities of women and Dalits (i.e. lower caste people) to access forest resources. The existing CF policies, legislation and guidelines of Nepal include various provisions so that local communities can have equitable use and access rights regarding forest resources. The forest operational plans of the case study CFUGs also include equity provisions related to the distribution of benefits and participation in CFUG decision-making processes. However, results of the case studies show that not all CFUGs are achieving this goal. Representation of women and Dalits on the Executive Committees of CFUGs, and their access to forest resources, is still lower than that of men and non-Dalits. Inequitable representation of women and Dalits is occurring not only in relation to the CF process, but also in relation to other community processes in the case study sites. Generally, there is lower participation of women and Dalits in political, economic and professional domains in Nepal (UNDP 2009). It can be argued that the representation of women and Dalits is lower not only because of low income, but also because of traditional social structures. Thus, careful consideration must be given to effective enforcement of gender and social equity provisions if poor people are to derive greater benefits from Nepal's CF program.

The majority (80 percent) of elite (rich and upper caste) members of the CFUGs interviewed claim that the handover of the forests to local communities has not only helped create a favourable environment for local involvement in forest protection, but that it has also helped to increase forest accessibility for poor and disadvantaged people, so that they can make use of forest products in accordance with the approved forest operational plans (FOPs). Elite members also reported that community forests are no longer fully controlled by the government or by elite and rich people, but are owned by members of the rural community, including poor and disadvantaged people, irrespective of their socio-economic backgrounds.

In interview the chairperson of the FECOFUN of Kabre Palanchok District said that the:

> initiation of community forestry has helped to create a favourable environment regarding the protection of forest resources by local people. It has also helped poor and disadvantaged people to have greater access to the forests, so that they can use forest products as per their approved forest operational plans.

In theory, CF exemplifies social justice, through equitable decision-making and benefit sharing; it does not discriminate on the basis of gender, caste, religion, interest groups, wealth or poverty. The participation of poor and marginalized members in decision making, benefit sharing, and conservation of forest resources processes is encouraged. Based on the perceptions expressed through in-depth interviews, equity in relation to community forestry could be defined as: 'the equal representation of marginal groups (especially women, poor people and Dalits) regarding decision-making, the distribution of forest products, and investment of CFUG funds'.

However, while 91 percent of household survey respondents believed that participation of women and poor in the protection and management of forests had been improving under CF, when asked about equity in decision-making and benefit sharing, 30 percent of all respondents said that there has been no improvement, and another 32 percent did not know whether there had been an improvement. Thus, 62 percent of the total respondents of the household survey had not recognized any improvement in outcomes under CF in relation to equity and social justice issues in their CFUGs (see Table 4.4a).

Table 4.4a Perceptions of equity in CF according to well-being as % of CFUG respondents (n=138)

CFUG	Improving (% of CFUG respondents)				Not improving (% of CFUG respondents)				Do not know (% of CFUG respondents)			
	poor	medium	rich	total	poor	medium	rich	total	Poor	Medium	Rich	total
KCFUG (n=50)	2	24	10	36	18	16	2	36	20	6	2	28
HJCFUG (n=35)	26	9	14	49	3	3	3	9	26	11	6	43
SDCFUG (n=53)	4	15	15	34	28	6	4	38	19	6	4	28
Overall	**9**	**17**	**13**	**38**	**18**	**9**	**3**	**30**	**21**	**7**	**4**	**32**

(Source: household survey 2012)

Table 4.4b Perceptions of equity in CF according to well-being status

Well-being	Do not know (%)	Improving (%)	Not improving (%)	n
Poor	44	18	38	66
Medium	22	51	27	45
Rich	19	67	15	27

In examining who is expressing particular views on equity, some clear disparities appear between groups. The perception of lack of improvement is more commonly expressed by poorer people, with 38 percent of all poor respondents indicating that equity is not improving, compared to only 15 percent of rich respondents. On the other hand, 38 percent of the respondents of the household survey believe that there has been an improvement in equity after the introduction of community forestry, and this view is more commonly expressed by richer members of the community. In this case, 67 percent of rich respondents believed that equity is improving, compared to only 18 percent of all poor respondents (Table 4.4b). Typically, the rich CFUG members believe that the poor are provided with enough forest products such as fuelwood for cooking, heating and lighting, and fodder for their cattle. Thus, the CFUGs have not been able to address the equity issue adequately through the implementation of the community forestry program, and this may be partly explained by the fact that wealthier CFUG members, who are more likely to hold executive positions, don't recognize equity issues in the same way that poor members do.

Interestingly, the poor were also much more likely to say that they do not know whether equity is improving (44 percent of all poor, compared to 22 percent of medium and 19 percent of rich respondents), which suggests the importance of social structures in people's willingness to express an opinion on a matter of social disadvantage (Table 4.4b). So, not only are the rich more likely to believe that the poor are adequately provided for, but the poor are less likely to speak up for themselves to draw attention to inequity.

Although natural resource management offers long-term benefits, it can create hardship for poor and marginalized people. Often the economic and cultural rights or interests of people are ignored by elite individuals, and even by government officials. During an in-depth interview, a former senior executive member of FECOFUN claimed that environmental conservation is not only an international issue, but that these natural resources are controlled by elite members of CFUGs, which makes it an issue for local resource users. In response to a question on equity and social justice in community forestry, this informant said:

> Illiterate people have been virtually blindfolded because of a lack of education and awareness; and this is why they are not able to stand up for themselves, regarding their community forestry rights. This is a misleading aspect of community forestry. People of all socio-economic backgrounds should be made aware of their rights and duties. Only then is it possible for equity and

social justice issues to be addressed, regarding decision making and benefit sharing.

During an interview, one respondent from Chitwan District Forest Office reported that one major difference in household economic status is that different members of the group have different needs for forest products. Although most households are largely dependent on subsistence farming, there are substantial differences regarding the landholdings of the richest and poorest:

> Poor people in reality do not need timber because they can't afford to build houses. What they need is small amounts of wood or bamboo for the construction of sheds; fuelwood for cooking and for keeping them warm in winter; fodder for their cattle; and fruits, medicinal plants and other non-timber forest products. In most cases, elite and rich people don't place a lot of attention on the development of non-timber forest products; but concentrate instead on the production of timber and fuelwood.

The average annual intake of forest products from community forest land varies from one CFUG to another; and from one wealth category of household to another (Table 4.5). The timber consumption habits of the wealthy households differ from those of poor households. Wealthy households use more timber for construction purposes than the poor and medium-income households, as rich people can spend money much more readily on construction than poor households can. So although the price of timber which is harvested from the community forests is low compared to the price in the local markets, poor households cannot afford to buy timber even at the lower price. As a result, most of the benefits from the low price of timber inadvertently go to the medium and rich households.

Table 4.5 Average annual household demand and supply of forest products from CF (n=138)

CFUG	Forest products	Poor	Medium	Rich
KCFUG	Timber (cft)	4.5 (4)	21 (17)	34 (25)
	Fuelwood (HL)	61 (55)	65 (60)	61 (43)
	Grass/forage	359 (332)	370 (296)	493 (271)
HJCFUG	Timber (cft)	13 (12)	29 (26)	37 (33)
	Fuelwood (HL)	84 (75)	122 (103)	120 (91)
	Grass/forage (HL)	238 (112)	300 (130)	386 (117)
SDCFUG	Timber (cft)	9 (8)	19 (16)	34 (28)
	Fuelwood (HL)	73 (71)	79 (77)	104 (91)
	Grass/forage (HL)	183 (124)	277 (169)	327 (155)

Note: cft: cubic feet; HL: head load (50 kg); first figure is mean annual demand of forest products per household, second figure in parentheses is supply from CFUG per HH.

(Source: household survey 2009)

The average annual demand for fuelwood and grass/forage from community forest land varies according to wealth categories of the households. Most wealthier and medium households source a large proportion of their fuelwood and grass/forage requirements from their own farmlands, while most poor people depend entirely on community forest land for their domestic need.

The Forest Act 1993 and Forest Regulations 1995 provide each CFUG with certain rights, on condition that the groups manage their local forest in accordance with government policies. However, despite granting local forest users certain rights, the legislation does not include any special provision regarding the specific needs of particular groups, such as the *Kamis* (the blacksmith artisan caste), whose livelihood depends on access to charcoal. Nonetheless, the Community Forestry Guidelines (DoF 2014) do contain special provisions regarding the supply of forest products to specific forestry based occupational groups, and such provisions may also be introduced into forest operational plans (FOPs) and group constitutions. Below (in Box 4.1) is the story of one *Kami* interview respondent from SDCFUG, where there is no special provision in the FOP regarding the supply of charcoal.

Box 4.1   Kami involvement in the rural community

A male respondent of the SDCFUG, of the '*Kami*' (or blacksmith) caste, a traditional forest-dependent occupational group, claims that he and other Kami people are playing a key role in relation to the economic development of the community by providing low-cost services and following traditional methods to make and repair agricultural tools that are indispensable for mountain farming.

Although the respondent is providing a vital service for the local people in relation to their livelihoods, he is one of a number of socially oppressed and economically disadvantaged members of the CFUG. The CFUG has introduced restrictions on the collection of wood for charcoal burning. If he had been given an opportunity to participate in the preparation of the FOP of the CFUG, then his need for fuelwood to produce charcoal may have been recognized in the FOP. However, because of his low socio-economic status in the community, he never got the chance to be involved.

Based on analysis of both household survey and in-depth interview responses, the government-managed forests of the past provided little in the way of socio-economic benefits to the local community and more than 90 percent of in-depth interview respondents believed that the implementation of CF has improved the biophysical condition of forests. Nonetheless, there is a recognition amongst many that CF programs have not achieved broad socio-economic benefits on the scale anticipated, especially in terms of poverty alleviation. In particular, issues about equity and social justice have not been addressed adequately, and equity and social justice provisions incorporated into CFUG constitutions and FOPs should be more effectively applied in order that the gap between poor and rich, and elites and general members, be reduced.

The CFUGs' memberships are not homogenous and there are diverging interests in regard to use of the forest. Conservation of forest is a priority of CFUGs, but individual interests vary when uses of products other than fuelwood and timber are in question. In all three CFUGs, wealthier households have greater use of forest products compared to poorer households, resulting in different incentives for forest use and forest management amongst different households. Wealthier households prefer forest products which support the agriculture system such as timber for building constructions, fuelwood and fodder, whereas poorer households prefer mostly small timber, dry fuelwood, leaf litter, and fodder which can generate cash and support their basic subsistence livelihoods. Many of the benefits of community forestry are flowing to local elites and poorer people are suffering from scarcity of forest products because of the inequitable distribution of forest products and other benefits.

In addition, patriarchal structures and processes such as gender division of labour and undervaluation of the contributions of women to support their household economy result in a low level of women's ability to attend meetings. Supplementary workload of poor and women members has also increased due to their engagement in forest protection and management activities. Because of their high workload and low socio-cultural position, their participation in decision-making and benefit-sharing processes is often inadequate. These groups of users have limited or no access to alternative resources such as trees on private land and little or no purchasing power for forest products from the local market. This situation has created a poverty trap and discrimination against this group of users.

Collection and mobilization of community funds

CFUGs receive significant revenue from the sale of forest products, as well as from other sources such as fines and micro-enterprises established by CFUGs (Table 4.6). The records of the case study CFUGs indicate that sales of products such as timber, green or large size wood, poles, bamboo, and non-timber forest products, provide most of the income. The records also show that the overall contribution by sales of forest products, to the three CFUG funds is, on average, approximately 81 percent of all money collected by the CFUGs. However, the exact percentage varies from one CFUG to another, with SDCFUG raising significant income from other sources (see also Chapter 6).

CFUG office records and audit reports also show how these funds have been utilized (Table 4.7). Across the three CFUGs, harvesting and logging of forest products is the greatest single cost reported in the records (33 percent), with a substantial amount re-invested back into the improvement and management of forests (17 percent). However, CFUGs can be seen to be making a substantial contribution to their local communities through investments in community

Table 4.6 Total income of the CFUGs from being handed over to 2012

Income sources	KCFUG		HJCFUG		SDCFUG		Total income of CFUGs	
	Income (NRs)	%	Income (NRs)	%	Income (NRs)	%	Income (NRs)	%
Forest products sales	34,050,129	84	6,279,902	97	3,597,116	51	43,927,147	81.5
Other incomes	6,310,274	16	211,714	3	3,426,910	49	9,948,898	18.5
Total (NRs)	40,360,403	100	6,491,616	100	7,024,026	100	53,876,045	100

(Source: data compilation and calculation from CFUG's office records, annual progress reports and audit reports 2012)

Table 4.7 Mobilization of CFUG funds from being handed over to 2012

Investment of CFUG funds, Nepal rupees (% of CFUG total spend)	KCFUG NRs (%)	HJCFUG NRs (%)	SDCFUG NRs (%)	Total amount	% of total expenditure of all CFUGs
Forest protection and development	7,144,334 (20.1)	928,415 (14.6)	351,839 (5.1)	8,424,588	17
Forest harvesting and logging	12,665.226 (35.5)	1,473,405 (23.1)	1,914,152 (27.6)	16,052,783	33
Income generating activities	1,021,338 (2.9)	0	1,914,097 (27.6)	2,935,435	6
Community development	7,321,470 (20.5)	3,255,760 (51.1)	1,172,757 (16.9)	11,749,987	24
Office management and administrative expenses	5,971,686 (16.8)	628,353 (9.9)	1,341,306 (19.3)	7,941,345	16
Miscellaneous	1,516,414 (4.3)	80,159 (1.3)	251,819 (3.6)	1,848,392	4
Total	**35,640,468**	**6,366,092**	**6,945,970**	**48,952,530**	**100**

(Source: data compilation and calculation from CFUG's office records, annual progress reports and audit reports, 2012)

infrastructure (24 percent) and income generating activities (IGAs) (6 percent). We can also see how the different CFUGs are adopting different strategies in the expenditure of their funds. KCFUG, for example, has a large area of community forest in the Terai where demand for Sal timber and other high-value products is high, and has a higher level of income generated from its forestry activities. KCFUG is investing a relatively high proportion (almost 56 percent) of income back into forest management, improvement and harvesting, and relatively little to support IGAs of individual households. SDCFUG, in the mountains, has a lower total income and spends relatively little on forest improvement and protection, but invests heavily in IGAs of households. HJCFUG, on the other hand, keeps its operational costs low, and has decided to return nothing back to individual households for IGAs, but to invest heavily in community development activities, such as school building, rural road, and drinking water projects (which may be more likely to deliver benefits to wealthier members than IGAs which are targeted to the poor). While the expenditure patterns may change from year to year in response to particular projects and needs, Table 4.7 shows clear difference in the strategies adopted by different CFUGs for expenditure of their funds, and thus the form in which benefits are realized by CFUG members.

Generally, IGAs especially for poor household groups in the villages involving cultivation of various types of grasses, fodders, mulberry for sericulture and fruit species could occur on forest land, if that land could be made available to poor people by CFUGs for such purposes. CFUGs could provide soft loans to individual households for various IGAs such as fruit and vegetable gardening, goat/pig farming, member shares for local micro-saving, and credit co-operatives and micro-enterprises. CFUGs also spent their funds for IGA activities at the community level such as establishment of a community-owned saw mill, rice and flour mills, shares for co-operative, micro-enterprises and ecotourism (e.g. swimming pool, botanical garden picnic spots and fish farming). CFUGs coordinate such programs with government departments and national and international non-governmental organizations (Devkota 2012). Data which is presented in Table 4.7 indicate that about six percent of total CFUG funds are spent on IGAs.

By comparing the total income from Table 4.6 with the total expenditures from Table 4.7, we can also see a different pattern. HJCFUG in the Middle Hills and SDCFUG in the Mountains spend only marginally less than their income for the year. However, with a much larger income from their natural Sal (*Shorea robusta*) forest with a high value of timber and non-timber products that provide a range of opportunities for use, KCFUG is able to retain almost 12 percent of their income. In SDCFUG and HJCFUG, forests consist of planted pine trees with relatively few other broad-leaved species and therefore less diversified use in an economic production system. These forests are mainly used for local consumption rather than for surplus income. In both CFUGs, there is a small market demand for pine timber from small local saw mills. These two forests are also located in relatively remote areas with steep topography leading to high costs of harvesting and transportation from forests to markets.

Key findings of the case studies

Improved forests have not delivered social equity

Prior to the introduction of the community forestry program, the government excluded local forest-dependent people from becoming involved in forest management. The government's 'command and control' forest management system was ineffective in the protection of forests. Forest resources were virtually unprotected. So, forest land was subjected to abuse, primarily because the government failed to enforce its policy in relation to the sustainable management of forest resources. Consequently, as reported by informants, the forests in all three study areas had deteriorated, with forest resources diminishing to the point where it became difficult for local rural people to meet their subsistence needs. This is consistent with the findings of other researchers (e.g. Nagendra and Gokhale 2008).

Informants in all three case study areas noted that the biophysical condition of forests has improved since the forests were handed over to local communities (i.e. CFUGs). Wildlife conservation in the community forests has been achieved by banning hunting and imposing heavy fines for the killing of wildlife. This improvement and the increase in the numbers of wildlife are regarded as being positive signs that biodiversity is being conserved in the community forests. This result is consistent with the findings of other researchers (e.g. Charnley and Poe 2007), who have concluded that local control over forest management has been successful in achieving positive ecological outcomes. These outcomes include the reduced rate of deforestation, increased forest cover and biodiversity, and maintenance of forest density.

The existing CF policy of the Government of Nepal is based on an assumption that the effective implementation of CF operation plans encourages the production of a variety of good-quality, economically valuable forest products (such as timber, fuelwood and fodder) as well as non-timber products such as wild mushrooms, fruits and medicinal plants (GoN 2015). However, the improved forests do not necessarily lead to improved social equity, and in some cases may actually be making life harder for some poor people (e.g. restrictions on charcoal production by *Kami*).

In theory, CF should exemplify social justice regarding decision-making and benefit sharing. Female and poor members of CFUGs should be encouraged and empowered to participate in the decision-making process. CF guidelines require that the CFUG Executive Committee should be comprised of at least 50 percent of members who are either female or poor (DoF 2014). Such an outcome was not seen in any of the case study CFUGs (Table 4.2). Although the existing forestry policies and legislation of Nepal include provisions for equitable use and access rights to forest resources, not all CFUGs are adequately enforcing those provisions. This study indicates that improvement in forest condition is easier and faster to achieve than is social transformation within rural communities, through the implementation of CF. Thus, achieving positive socio-economic outcomes is challenging. Carefully designed and targeted policies need to be implemented, as without such policies, elite capture of the benefits from CF are reinforced.

In relation to CF, there is still a lack of social equity regarding CFUG committee membership, decision-making and benefit sharing. This is occurring because of the power disparity between elite and disadvantaged users, an inadequate supply of information at the local level regarding policy, and a lack of interest amongst some DFO and forestry field staff (Devkota 2012). Office records of the CFUGs in the study areas indicate that the Executive Committees, which have been in operation since the first year the forests were handed over, are dominated by high caste wealthy men and medium-wealth, upper caste men and women (Table 4.2). The results of the case studies indicate that the participation of women, poor people and disadvantaged groups on Executive Committees and in the decision-making process is very low, while local elites (i.e. people who are wealthy, well-educated and of high social status) are influential and dominate CF processes.

Socio-economic characteristics of CFUGs are still influenced by traditional beliefs in Nepal. In all three of the study area communities, rich residents, who receive regular amounts of cash income, are often upper or middle caste, and dominate socio-economic life. The subsistence concerns of the minority and marginalized groups are easily overlooked. The caste system, comprised of unequal layers in a form of social stratification, suppresses lower caste users regarding access to, and control over, forest resources. The majority of lower caste (Dalit) members are socially oppressed and economically poor. As women and poor and disadvantaged members are largely excluded from the CFUG decision-making process, the community forestry program has not been very successful in improving social equity amongst forest users. Therefore, improved forest conditions can be achieved without improved social equity within local communities.

Improved forests have not delivered improved economic equity

Although the biophysical condition of community forests is improving because of CFUG management, the CF program still faces organizational, structural and societal challenges. The case study indicates that the occupants of rich and middle-income households benefit most as a result of the CF program.

Results of the case studies show that economic wealth and social status are inextricably linked in these rural communities. The educated wealthy have higher status and are more likely to hold decision-making positions within the CFUGs. High-income households also receive the greatest share of benefits from timber of CF (Table 4.5), followed by the medium-rich class. The occupants of rich and middle-income households can afford to buy timber and other forest products. Poor people cannot afford to buy timber and firewood, even at the subsidized rate; therefore such a rule introduced by the Executive Committee of the CFUG actually restricts access to fuelwood for poor people.

Analysis of the selected case studies found that there has been an overall improvement in the economic condition of CFUGs. For example, not only have informant interviews and CFUG records indicated an increase in the quantity of forest products being acquired and used, but there has also been an increase in the quality of those items. But improvement has not necessarily occurred in relation to the socio-economic condition of all individuals, especially poor and disadvantaged families. Indeed, economic data aggregated at the community level can mask the disparity of economic and social benefits received by individual

families. As product distribution decisions are made by elites and powerful people, poor people and Dalits often lack access to forest products, and thus do not
receive economic benefits from community forestry to the same extent that well-
off residents do. The inability of certain people to participate in the benefit-
sharing mechanism often leads to disinterest regarding participation. The data
presented in Table 4.5 suggests that occupants of medium-income households
benefit the most from timber, fuelwood and grass/forage of community forestry.

Poor CFUG members, on the other hand, do not have enough trees on their
private land, and consequently rely heavily on community forests. They either
have to wait for the forest harvesting times, which are set by the CFUG ECs, or
steal forest products at night, in contravention of the FOPs. The restriction in
supply of forest products presents problems for those who depend most heavily on
those forest products, resulting in dissatisfaction with the Executive Committee,
and frustration at the continued lack of meaningful representation of poor and
Dalit people on the Executive Committee. Thus, the CF program has not yet
been successful in addressing some of the more profound socio-economic equity
issues facing poor people in rural areas.

Community forestry may be exacerbating poverty in some households

Some of the results from the case studies indicate that CF is a 'user group-focused'
program, rather than a 'poverty reduction' program, because rich people cannot
be excluded from the use of forest resources. Several factors such as different
interests of users, heterogeneous ethnic composition, conflicting political beliefs,
different economic condition and different socio-cultural backgrounds within
each community can create problems for the equitable distribution of benefits to
community forestry.

This mirrors Pokharel's (2008) results, which indicate that 74 percent of the
total benefits that result from community forestry in Nepal are received by the
non-poor, while only 26 percent are received by the poor members of rural communities in Nepal. Gilmour, Malla and Nurse (2004) have presented empirical
evidence that CF can provide tangible economic benefits for rural communities.
However, the authors also note that there is no clear evidence that CF has substantially increased the benefits for poor people, despite the huge potential that
CF may have regarding achieving pro-poor outcomes.

The main forest-related issues facing poor people in Nepal are: the lack of access
or limited access to fodder and fuelwood for daily requirements, the increased
burden associated with forest management, the lack of purchasing power and
their lack of influence in decisions regarding forest use (Devkota 2012). These
issues may be occurring because of constraints such as lack of time, no access to
private land resources, low level of education, low level of awareness, lack of
confidence and limited power in institutional processes. To overcome structural
inequities, the CFUG FOPs include provisions to make cash loans to poor people
for the establishment of micro-enterprises controlled by women and poor people,

and for areas of community forest to be set aside for fodder collection. However, the case study findings indicate that elite members of the CFUG Executive Committees (CFUG ECs) are reluctant to put such provisions into actual practice. DFO staff provide limited monitoring of CFUG EC activities, and are therefore unable to influence elites in relation to the development of pro-poor practices.

The case study results indicate that elite people are more able to influence decisions and to reap the benefits. CFUG EC members hold the power to make decisions in relation to use of forest products, collection and mobilization of their community fund, and what is to be discussed at meetings and in annual general assemblies. As a result, the concerns of women, poor people and Dalits, who depend more on forest resources for their livelihoods and are excluded from the decision-making and benefit-sharing processes, are not being properly addressed (Devkota 2012). With limited capacity to generate alternative income from private lands or from the market, the basic needs of poor people are not being met. Rather than solving the poverty issues, CF can be considered to be reinforcing the poverty condition of the poorest members of the CFUGs. Thus, contributions of CF to reduce the household poverty of individual poor and Dalit households are not only insufficient, but may be exacerbating the poverty situation of some households.

The findings presented in this chapter are consistent with that presented by Malla (2000), with his analysis of the socio-economic impact of CF in Nepal revealing that fixing a price and quota for forest products for users purchasing them from the CFUG disadvantages poor users. They are entitled to a share, but simply cannot afford to pay for it. They are thus often worse off than before CF was introduced. Shrestha (2005, 2016) has argued that having strict rules of equality of costs and access can lead, paradoxically, to inequitable results, because the poor need more from community forests than wealthier people.

Protection-oriented forest management does not necessarily produce livelihood benefits

The Nepal government policy in relation to CF is directed not only at protecting and rehabilitating existing forests, but also at improving the livelihoods of people in local communities through the wise use of community forest resources. However, in reality these multiple objectives are not being met. A review of the FOPs of the CFUGs in the three case studies and extensive interviews and surveys indicate that the CFUGs are 'protecting trees' rather than 'managing forests'. The FOPs are still largely protection oriented and are conservative prescriptions regarding the types and quantities of forest products that local people can harvest. Little in the way of attention is given to silviculture and other management aspects (Devkota 2012). This produces a situation whereby there is a minimum flow of forest products to user members.

CFUGs have imposed restrictions on the harvesting and selling of forest products (Devkota 2012). The harvesting prescriptions in the CFUG operation plans have been found to be very conservative. Although the quality and quantity of forest products have increased because forests are being protected by local communities, CFUGs have not allowed the harvest of sufficient quantities of forest products to meet the basic needs of all community members. Therefore, the average annual quantity of forest products harvested from community forests is less than the annual allowable harvest (AAH) of forest products. Because of the implementation of conservative FOPs, the harvesting and thinning of trees are being delayed. As a result, forest land is becoming dense, rather than being properly managed. This situation does not create an enabling environment to produce optimum socio-economic benefits of CF to support the livelihoods of the rural poor.

Gaps in the implementation of community forestry

The government's assumption that frames CF is that intervention is necessary not only in order that the subsistence livelihoods of the poorest forest users be improved, but also so that opportunities can be provided for the poor to lift themselves out of poverty through the use of forest product micro-enterprises. These could perhaps be developed and managed by poor members of CFUGs on an entrepreneurial basis. However, local elites have been allowed to capture many of the benefits from CF at the expense of poor members (Devkota 2012).

The policy guiding the CF program in Nepal indicates an expectation that the implementation of CF will result in forest management becoming socially inclusive and capable of enhancing human livelihoods, natural resources, and the rights of poor people and those who are socially excluded. There is a provision in the existing forestry policy whereby some part of local community forest land could be set aside for occupants of poor households so that NTFPs could be cultivated, as well as other provisions that support pro-poor income generating schemes (DoF 2014). However, findings from the case studies indicate that such provisions in the existing CF guidelines have not been put into practice. While pro-poor provisions have been included in CFUG FOPs, there has been little progress regarding the utilization of these provisions for the well-being of the poorest community forest users. As long as such provisions are not implemented effectively by CFUGs, it is unlikely that community forests will be able to provide for the subsistence needs of the poorest in the community. Pro-poor activities constitute only a small proportion of CFUG efforts. Thus, poor people and Dalits are receiving limited benefits from CF.

Many challenges remain regarding equity in participation and the sharing of benefits from CF. For example, the existing CF guidelines contain provisions whereby affirmative action can take place on behalf of women, poor people and occupants of marginalized households. The guidelines include a provision whereby at least 35 percent of total CFUG income can be allocated for the improvement of livelihoods of occupants of poor households, based on an assessment of well-being. As shown in Table 4.7, different CFUGs have adopted different priorities in how they spend their money, but across the three case study CFUGs, only six percent of total income has been spent on income generating activities and pro-poor programs. Thus, the effective implementation of the existing provisions of CF guidelines for improvement of livelihood of poor households is an urgent need and challenge to address socio-economic inequities and to enhance the capacity of local institutions to act collectively.

Conclusions

The emergence of CF in Nepal resulted from recognition of the failure of the previous government-managed forestry policy, which was ineffective primarily because it did not involve local people in the management and development of forest resources. The government-centred forestry policy – characterized by ignorance of local people's potential management roles, needs and user rights – has been replaced by a local-level, elite-led approach that continues to largely ignore the rights and needs of marginalized people, and the role that they should play in CF.

The exploitative behaviour of upper caste male elites dominates the three case study CFUGs. CF exists within a dynamic environment, having to respond to changing environmental, political and socio-economic conditions in order to deliver socio-economic outcomes to local communities. Just a handful of elite community members occupy the key decision-making positions on Executive Committees. In this context, CF cannot reach its full potential without adapting quickly and effectively to external changes. The reinforcement of traditional power structures within community forestry is a major obstacle to achieving optimum socio-economic benefits from CF for rural livelihood improvement and poverty reduction. It is difficult for women, poor people, and disadvantaged individuals to participate actively in the decision-making and benefit-sharing processes. This exclusion from positions of power or influence is based on existing social hierarchy, reinforced by the structures of CF. Furthermore, the domination of CFUG committees and the decision-making process by elites and rich members leads to a lack of acknowledgement of the needs and interests of poor people, women and disadvantaged individuals. Unless the CF policy is reformulated and effectively enforced to deliver clear socio-economic benefits to the most marginalized in rural communities, it will struggle to achieve meaningful outcomes against its rural livelihoods and poverty reduction objectives.

In conclusion, our case study research has shown that there have been demonstrable improvements in forest condition, but not necessarily improvements in the social and financial well-being of all CFUG members; specifically, the poorest forest users may be left worse off as a result of management decisions and practices of their CFUG Executive Committees. Biophysical outcomes can be misleading indicators of the success of community forestry, and do not reflect social and economic outcomes. Investment focussed solely on achieving biophysical change can be misguided and can entrench social inequity at the local level.

This chapter has considered issues related to socio-economic outcomes of CF, particularly in relation to those aspects that are directly related to access to, and use of, forest resources, and the collection and mobilization of CFUG funds for improvement of rural livelihoods and poverty reduction. A range of other potential socio-economic outcomes of community forestry such as changes in skills and knowledge amongst the local community, literacy status, leadership development, gender empowerment, social capital formation and infrastructure development are considered as components of a broader concept of community development in Chapter Six.

References

Acharya, K.P. and Acharya, S. 2007, Small scale wood based enterprises in community forestry: Contribution to poverty reduction, *Banko Janakari*, vol. 17, no. 1, pp. 3–10.

Adhikari, B. 2003, *Property rights and natural resources: socio-economic heterogeneity and distributional implications of common property resource management*, South Asian Network for Development and Environmental Economics (SANDEE), Kathmandu, Nepal.

Adhikari, B., Falco, S.D. and Lovett, J.C. 2004, Household characteristics and forest dependency: Evidence from community property forest management in Nepal, *Ecological Economics*, vol. 48, no. 2, pp. 245–257.

Agarwal, B. 2009, Rule-making in community forestry institutions: The difference women make, *Ecological Economics*, vol. 68, no. 8–9, pp. 2296–2308.

Bhattarai, B. 2016, Community Forest and Forest Management in Nepal, *American Journal of Environmental Protection*, vol. 4, no. 3, pp. 79–91.

Bhattarai, S., Jha, P.K. and Chapagain, N. 2009, Pro-poor institutions: Creating exclusive rights to the poor groups in community forest management, *Journal of Forest and Livelihood*, vol. 8, no. 2, pp. 1–15.

CBS 2007, *Population Profile of Nepal*, Central Bureau of Statistics, Kathmandu, Nepal.

CBS 2012, *National Population and Housing Census 2011 (National report)*, National Planning Commission Secretariat, Central Bureau of Statistics, Government of Nepal, Kathmandu.

Chambers, R. and Conway, G.R. 1992, *Sustainable rural livelihoods: Practical concepts for the 21st century*, IDS Discussion Paper No. 296, Institute of Development Studies (IDS), Brighton, UK.

Charnley, S. and Poe, M.R. 2007, Community forestry in theory and practice: Where are we now? *Annual Review of Anthropology*, vol. 36, pp. 301–336.

Devkota, B.P. 2012, *Socio-economic outcomes of community forestry in Nepal: Lessons from three diverse rural communities*, PhD thesis, Charles Sturt University, viewed 22 February 2017, http://primo.unilinc.edu.au/primo_library/libweb/action/dlDisplay.do?vid=CSU2&docId=dtl_csu40107.

Devkota, S.R. 2005, Is strong sustainability operational? An example from Nepal. *Sustainable Development*, vol. 13, pp. 297–310.

DFID 2001, *Sustainable Livelihoods Guidance Sheets*, Department for International Development, London, UK.

DoF 2014, *Community Forestry Development Program Guidelines*, 3rd edn, Department of Forests, Government of Nepal, Kathmandu.

FAO 2010, *Global Forest Resource Assessment 2010 (Main Report)*, Food and Agriculture Organization of the United Nations, Rome, Italy.

Fisher, R.J. 2014, *Lessons Learned From Community Forestry in Asia and Their Relevance for REDD+*, USAID-supported Forest Carbon, Markets and Communities (FCMC) Program, Washington, DC, viewed 26 February 2017, www.fcmcglobal.org/documents/CF_Asia.pdf

Gilmour, D., Malla, Y. and Nurse, M. 2004, *Linkages Between Community Forestry and Poverty*, Regional Community Forestry Training Center for Asia and the Pacific (RECOFT), Bangkok, Thailand.

GoN 1989, *The Master Plan for the Forestry Sector*, Ministry of Forests and Soil Conservation, Government of Nepal, Kathmandu, Nepal.

GoN 2015, *Forest Policy, 2015*, Ministry of Forests and Soil Conservation, Government of Nepal, Kathmandu, Nepal.

Kanel, K.R. 2004, Twenty five years of community forestry: Contribution to millennium development goals, in R.K. Kanel, P. Mathema, B.R. Kandel, D.R. Niraula, A.R. Sharma and M. Gautam (eds), *Twenty-Five Years of Community Forestry*, Proceedings of the Fourth National Workshop on Community Forestry 4–6 August 2004, Kathmandu, Nepal, pp. 4–18.

Kanel, K.R. and Dahal, G.R. 2008, Community forestry policy and its economic implications: An experience from Nepal, *International Journal of Social Forestry*, vol. 1, no. 1, pp. 50–60.

Maharjan, K.L. and Khatri-Chhetri, A. 2006, Role of forest in household food security: Evidence from rural areas in Nepal, *Annual Report of Research Center for Regional Geography*, vol. 15, pp. 41–67.

Malla, Y.B. 2000, Impact of community forestry on rural livelihoods and food security, *Unasylva*, vol. 202, no. 52, pp. 37–45.

Malla, Y.B. 2009, Community forestry: Past, present and thoughts for the future, *Proceedings of the community forestry international workshop 'Thinking globally-acting locally: Community forestry in the international arena*, 15–18 September 2009, Pokhara, Nepal.

Malla, Y.B., Neupane, H.R. and Branney, P.J. 2003, Why aren't poor people benefitting more from community forestry? *Journal of Forest and Livelihood*, vol. 3, no. 1, pp. 78–93.

MFSC 2008, *The future of Nepal's forests: Outlook for 2020*, Asia Forestry Outlook Study 2020: Country Report NEPAL, submitted by Ministry of Forestry and Soil Conservation, Kathmandu, Nepal, to Food and Agriculture Organization of the United Nations, Bangkok, Thailand.

Nagendra, H. and Gokhale, Y. 2008, Management regimes, property rights and forest biodiversity in Nepal and India, *Environmental Management*, vol. 41, no. 5, pp. 719–733.

Pokharel, R.K. 2008, *Nepal's community forestry funds: Do they benefit the poor?* SANDEE Working Papers No. 31–08, SANDEE, Kathmandu, Nepal.

Shrestha, K.K. 2005, *Collective action and equity in Nepalese community forestry*, PhD thesis, University of Sydney, Sydney, Australia.

Shrestha, K.K. 2016, *Dilemmas of Justice: Collective Action and Equity in Nepal's Community Forestry*, Adroit Publishers, New Delhi, India.

UNDP (United Nations Development Programme) 2009, *Nepal Human Development Report 2009: State Transformation and Human Development*, United Nations Development Programme, Rome, Italy.

World Bank 2008, *Forests Sourcebook: Practical Guidance for Sustaining Forests in Development Cooperation*, World Bank, Washington, DC.

5 Community forestry and community development in Nepal

Binod Devkota, Richard Thwaites and Digby Race

Introduction

This chapter considers the role of community forestry (CF) as an agent for social mobilization that leads to community development. CF institutions have been established throughout the rural communities of Nepal, introducing a level of local autonomy and responsibility not previously experienced, and offering opportunities for building skills and capacities. This chapter considers how the enhanced human and social capitals have contributed to positive social change in many rural communities, and how CF has provided a central institutional focus for the community, as well as for external agencies seeking to facilitate community development. Examples are provided of communities seeking to expand the agenda of their local CF group so that a broader local 'development' agenda is considered, and the directions that this 'new' agenda have taken, including micro-finance and alternative livelihoods.

Definition and concept of community development

Community development combines the ideas of 'community' and 'development' (Pawar 2010). Literature on 'community' shows that the definition of community varies in different contexts; as such it can be based on a way of life, geographical location, social system (including norms and values), and power structures and relationships (Cavaye 2006; Pawar 2010). A classical definition of community is that it is a group of people who live in a specific geographical location with some common interests or issues (Pawar 2010). DeFilippis and Saegert (2012) believe that the essence of a community is based on people having a shared place to live and work, though not necessarily doing both in the same place. However, in this chapter, we adopt a definition of community that covers a wide group of people that are located in a specific geographical region and are dependent on forests for their income, livelihoods and well-being.

In this case, they may be more or less dependent on forests. The community of place would be those people living in the vicinity of the forest. Within that 'community', there may be a range of other communities – communities of identity (e.g. females, Dalits), communities of practice (e.g. people with different forest-based enterprises or livelihoods) – who would then have quite different needs from forests and put different demands on forest resources.

Community development is a process that provides opportunities for, or adds value to, the lives of people belonging to the designated community (Gilchrist and Taylor 2016; Pawar 2010). The community itself takes action with a degree of coordinated or uniform participation. In this process, not only are local people creating more jobs, income and infrastructure, but they can also help their community to improve their ability to manage social change (Cavaye 2006; DeFilippis and Saegert 2012). Community development seeks to empower individuals and groups of people by providing the necessary skills and knowledge for people to effect change in their own communities. These skills and knowledge often focus on building political power through the creation of social groups working towards a common agenda. A holistic view of community development is that it is a process that supports people in taking collective action to build communities based on justice, equality and mutual respect and also to solve existing problems and improve livelihoods (Gilchrist and Taylor 2016).

A holistic approach to community development often includes elements of economic development, such as meaningful community participation, building skills and creating knowledge. Yet economic development often focuses on increasing employment, income generation and expanding the community's economic base. However, at times, economic development may not lead to community development, especially if the benefits of increased economic activity are enjoyed by a small segment of the community at the expense of others. This chapter explores the extent CF has led to community development in rural Nepal.

Community development through community forestry

According to the Forest Act 1993 of Nepal, out of the total annual income earned by a CFUG from the implementation of its forest operational plan, at least 25 percent should be used for the development, protection and management of community forests (GoN 1993). The central role of CF is to manage forests for improved forest regeneration and condition; thus to increase productivity of forest areas and improve livelihoods of the surrounding rural community. As forests mature, it is anticipated that a wider range of resources and services to the local community will ensue. The core socio-economic outcome of CF is the provision of forest products for livelihood needs (timber, fuelwood, fodder, NTFPs, as discussed in Chapter 4), but CF is able to contribute to communities in many ways other than just by generating products and income. A key strategy for CF to support community development is by generating funds that are then used for the development of human and social capitals of CFUG members. Beyond the 25 percent of funds designated for forest management, the guidelines prepared in 2014 for the Community Forestry Development Program have a mandatory provision that 35 percent of the total annual income should be used for activities targeted to the poor (women, Dalit, indigenous people and ethnic groups) as identified from participatory well-being ranking (DoF 2014). The remaining

40 percent of CFUG income should be spent on office management and other community development works (e.g. providing drinking water facilities, bridge and road construction, school improvements) as per the annual plan approved by the general assembly.

The holistic understanding of the concept of 'community development' used in this chapter includes local industry development. In other words, community development is an umbrella concept that integrates all aspects of the social, economic and ecological environment in which rural livelihoods exist. In this context, development is defined as the process that increases choices for members of the CFUG and delivers an improvement in overall rural community conditions, incorporating economic and other quality-of-life considerations such as environment, health, infrastructure and housing. Because rural poverty is a multidimensional problem, outcomes of CF have the potential to contribute to a wide variety of community development activities that support livelihoods. So, based on this understanding, we use the term community development to incorporate all the activities and benefits that contribute to the enhancement of the community apart from the core role of CFs in supporting forest regeneration and provision of forest products (Figure 5.1).

The initial concept of CF was that there be sustainable use of resources for subsistence needs. The concept gradually began to relate to community development, social safety nets and helping the poor to combat poverty (Mahanty et al. 2006; NSCFP 2007). Researchers, such as Springate-Baginski and Blaikie (2003), regard community forestry in Nepal as being not just a one-time policy change, but also an ongoing and evolving socio-economic development process. The engagement of a CFUG in community development activities is regarded as being one of the indicators of a good CFUG (Devkota 2012). It is believed that

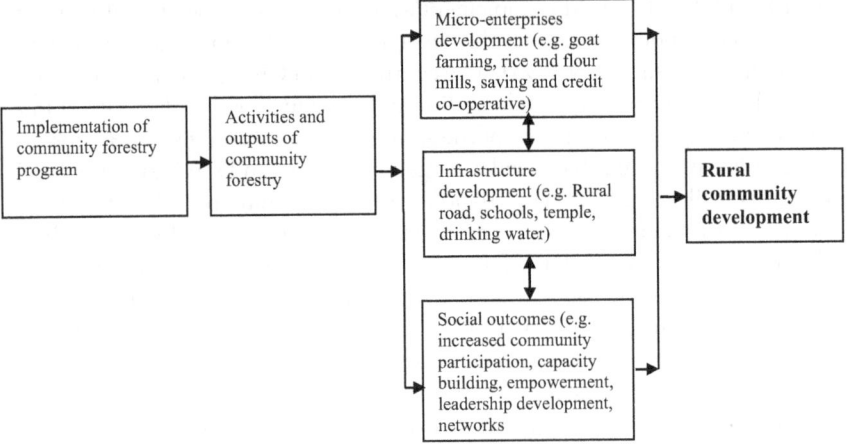

Figure 5.1 Outcomes of community forestry for rural community development
(Source: Devkota 2012)

as CFUGs carry out their basic operations successfully, they often move towards wider community development activities such as the supply of drinking water, irrigation, and the building of health posts, schools and roads (Adhikari and Adhikari 2010; Yadav 2004; Springate-Baginski and Blaikie 2003).

Many CFUGs are now moving beyond merely meeting subsistence needs and are considering a market-oriented scheme for their forest products and multiple uses of their forest resources (Chhetri 2006; Devkota 2012). The community forestry program guidelines propose that all community development activities should reflect the interests and needs of the rural communities concerned, including the needs and interests of poor people, women and disadvantaged members of CFUGs (DoF 2014).

This chapter presents case studies of three CFUGs from the Chitwan (Kankali CFUG or KCFUG), Kabre Palanchok (Hile Jaljale CFUG or HJCFUG) and Sindhu Palchok (Shreechhap Deurali CFUG or SDCFUG) Districts of Nepal, as described in greater detail in Chapter 4 of this book, and derived from the detailed PhD research of Devkota (2012). The data is drawn from forest operational plans (FOPs), constitutions, office records of the CFUGs, field observations, household surveys and focus-group discussions, as well as in-depth interviews with CFUG Executive Committee members, CFUG members, FECOFUN representatives, community forestry experts and government forestry officials.

Devkota's (2012) research found common contributions of CF to community development include:

- Micro-enterprise development (e.g. micro-finance for goat farming, rice and flour mills, saving and credit co-operatives);
- Infrastructure development (e.g. bridges and roads, schools, temples, pipes and tanks for drinking water); and
- Social outcomes (e.g. increased community participation, capacity building, empowerment, leadership development, expanded and stronger social networks).

Investment in household and community income generating activities through community forestry

Forest land could potentially be used for income generating activities (IGAs), such as the cultivation of various types of grasses, fodder, mulberry, medicinal species and fruit species (Yadav 2004; Devkota 2012). The CF guidelines (DoF 2014) make allowance for degraded forest land to be made available to poor households for cultivation of forestry-related cash crops as income generating activities. CFUGs also provide grants and soft loans for households to develop their own income generating activities, such as goat farming, or supporting households to shift from growing only subsistence food crops to growing vegetables as a cash crop for sale in the markets. CFUGs can also invest in co-operatives and income generating activities at a community level. Examples of such investments are provided in Table 5.1, some of which are further discussed below.

Table 5.1 Allocation of CFUG funds for income generation and community development (in period from handover of community forest to 2012, expenditure figures in Nepalese rupees [NRs])

Activities	Kankali CFUG	Hile Jaljale CFUG	Shreechhap Deurali CFUG	Total of all CFUGs	% of total expenditure of CFUGs
Years of data from handover of forest to 2012	17 years	14 years	14 years		
IGA for poor households					
Goat/pig farming	210,831	0	73,837	284,668	0.6
Soft loan for vegetable gardens	45,743	0	193,269	239,012	0.5
Fodder/ forage/fruit/ broom grass production	41,456	0	0	414,560	0.1
Sericulture	40,263	0	0	40,263	0.1
Investments for co-operatives and micro-enterprises					
Investment in co-operatives	0	0	133,100	133,100	0.3
Investment in micro-hydropower	0	0	121,110	121,110	0.2
Fish farming	18,150	0	0	18,150	0.01
Saw mill	0	0	247,588	247,588	0.5
Rice and flour mill	0	0	53,240	53,240	0.1
Truck	0	0	83,5653	83,5653	1.7
Lapsi candy	0	0	25,6300	25,6300	0.5
Ecotourism development					
Botanical garden	43,258	0	0	43,258	0.1
Swimming pool	440,000	0	0	440,000	0.9
Picnic spots	181,637	0	0	181,637	0.4
Total	**1,021,338**	**0**	**1,914,097**	**2,935,435**	**6**

(Source: data compilation and calculation from CFUG's office records, annual progress reports and audit reports)

Goat farming

According to office records of the case study CFUGs, CFUG funds have supported the establishment of goat farming as an income generating activity for poor and disadvantaged CFUG members. However, not all of the allocated funds have actually been spent in practice. For example, a total allocation over a 17-year period by KCFUG was NRs 210,831 for supporting goat farming and piggery in about 40 poor and very poor households. However, over this period, only

NRs 22,500 (approximately USD350 at 2017 exchange rate) had actually been provided in the form of loans to about 10 of those households to buy goats. The main reason for not spending the allocated budget for goat and piggery farming was the lack of commitment and confidence of the Executive Committee in making what was considered to be a risky investment to poor members. Under the loan arrangements, the money (or the equivalent in goats) must be paid back to KCFUG after two years, and that money will then be loaned to other poor CFUG members. However, the research identified problems in the design of these soft loan schemes. Some poor recipients had no experience in handling goats, and with little support available, were unable to make a success of goat farming. Of the 10 households that had received support for goat farming, only three have permanently adopted goat farming as an IGA. Group discussions in KCFUG revealed that loan periods of one or two years for income generating activities such as goat farming and vegetable farming are too short, and that loan recipients cannot earn enough income from the new activity to repay their loan within the time period. To pay back the loans to the CFUG, households have had to collect fuelwood and sell it in the local market.

Investments in co-operative and micro-enterprise development

The CFUG remains the central institution for running co-operatives (collection of funds from individuals and CFUG) to establish saving and credit micro-finance and micro-enterprises. The Forest Act 1993 and Forest Regulations 1995 recognize CFUGs as being non-profit, non-government organizations (NGOs). There is a provision regarding the use of a CFUG fund to allocate budget as shares in a co-operative to poor member households. So, CFUGs can invest funds in the establishment of co-operatives and micro-enterprises (i.e. forestry- and agriculture-based small businesses established either by individuals or by the local forest user group) to provide soft loans and employment respectively for poor user households. The fund can be used for both poverty reduction programs and local community development activities, as decided by the CFUG general assembly. For example, SDCFUG has been particularly active in investing money in the establishment and development of community-owned micro-enterprises, in the hope that employment opportunities for CFUG members would be increased and rural poverty would be reduced. CFUGs can also use their funds for establishment and development of co-operatives for investment in income generating and community development activities. SDCFUG has spent NRs 1,100,000 from its own fund towards establishment of the *Social Transformation Saving and Credit Cooperative*, and this cooperative has in turn invested about NRs 100,000 (about USD1,500 at 2017 exchange rate) as a soft loan for establishment of a *lapsi* (*Choerospondias axillaris*) candy business established by the SDCFUG as a local micro-enterprise. The co-operative and micro-enterprise are both managed and operated by members of the SDCFUG Executive Committee. Thus, community-owned micro-enterprises can provide employment for CFUG members whereas co-operatives are local financial institutions to provide soft loans to shareholders for income generating activities (e.g. vegetable farming, goat farming).

Lapsi is one of the most popular fruits in Nepal. It is grown mostly on private land and in community forests, and is used by local people as a cash crop. There is a high demand from candy factories for *lapsi* fruit, both inside and outside the district. Farmers who grow *lapsi* trees can obtain cash advances from NRs 500 to NRs 1,000, depending on the number of trees and the amount of *lapsi* fruit likely to be harvested. Farmers seeking an alternative source of income will often plant *lapsi* trees on their private land. SDCFUG supports this activity by providing free seedlings and assuring the farmers that their fruit will be purchased for use in its candy factory. The candy enterprise provides income for some poor people including through employment of about 40 poor women in candy production. However, this newly established forest-based micro-enterprise requires support from people outside the community for the proper marketing of its candy products so that the business can provide optimum benefits in a sustainable form to CFUG members.

SDCFUG has also invested money in the establishment of a saw mill, with technical assistance from the Nepal-Australia Community Resource Management Project. This mill can add value to locally harvested timber by converting logs into sawn timber for the local markets. This mill requires a highly skilled manager and some trained people for efficient operation. As the SDCFUG does not have the capacity to run this mill, it has been leased to a contractor (a businessman who is not a member of the CFUG) for NRs 64,000 per year. This provides a regular annual income source to the CFUG, while the remaining profit from this mill goes to the contractor. Under the lease agreement, SDCFUG has secured a provision whereby it provides a subsidy for the operation of the mill and in return SDCFUG members can obtain sawn timber from the mill at a discount rate to local market prices.

SDCFUG has also provided money for the establishment of a rice and flour mill. This SDCFUG-owned mill provides rice and flour at a cheaper rate than the cost offered by outside mills, which is especially appreciated by poor and disadvantaged CFUG members.

SDCFUG also spent NRs 1,100,000 to purchase a truck so that logs and other goods can be easily transported. This truck has often been used by both the local people and the saw mill contractor; however, at the time of research data collection, the truck was leased by SDCFUG to an elite member of the SDCFUG for NRs 20,000 per month. As well as this regular income going into the CFUG fund, two poor and marginalized members of the SDCFUG are employed as driver and helper to operate this truck. From the information obtained from the CFUG, it is unclear how much of the benefit is being captured by elite and wealthier members of the CFUG. As well as the business opportunity going to the elite member holding the lease, the truck is available to transport agricultural farm products and other goods at a discounted or subsidized price, a benefit experienced mostly by rich and medium-income member households. The question arises whether benefits arising from the investment in a truck might be more evenly distributed if the CFUG had retained the right to operate the truck in the interests of all its members.

While SDCFUG has been the most active in investing in a co-operative and micro-enterprises, other CFUGs are supporting income generating opportunities in different ways. For example, HJCFUG has established, with the help of the Micro-Enterprise Development Program (a national program for establishment and development of micro-enterprises in Nepal), a workshop for the *Kami* (or blacksmith caste) occupants of 14 households for improved manufacturing of agricultural tools and equipment used by local farmers. Furthermore, HJCFUG has also provided training for Dalit people in bio-briquette production and marketing, while the Micro-Enterprise Development Program (MEDEP) has supported the establishment of facilities for bio-briquette production, and the marketing of the bio-briquettes in Banepa and Kathmandu. These micro-enterprises actually contribute to generating employment for poor and Dalit households and quality service delivery to the local community.

Ecotourism development

Ecotourism could provide a potential source of income for local communities derived from people's desire to visit and enjoy the natural environment offered by community forests. Adopting an ecotourism approach could also ensure that the tourism activities are sustainable by managing the impacts on the forest and ensuring visitors have positive experiences that engage them with the economic, environmental and cultural values of the forest for local people (Bhattarai 2011). Thus, ecotourism could play an important role in not only economic growth of rural communities, but also in forest resource conservation. However, there is no understanding in the local villages of the nature of ecotourism, and no efforts to manage tourism impacts or engage visitors with an understanding of the forest.

Ecotourism does offer a potential source of income for CFUGs located near urban and tourist areas. KCFUG and HJCFUG are located near tourist centres, namely Chitwan National Park and Nagarkot, respectively. Although there is provision for each of those CFUGs to invest money in tourism, only KCFUG has spent money on the establishment of picnic spots, a botanical garden, a viewing tower, a swimming pool, a trekking trail, and a children's playground inside the community forest area, with the expectation of obtaining additional income as a result of those investments. These activities help to create additional employment for poor and disadvantaged members of the CFUG. KCFUG also charges fees for use of these facilities and all incomes from these activities go to the CFUG fund. However, planning and implementation of these ecotourism activities have been on an *ad hoc* basis. KCFUG has not developed a business plan or undertaken any kind of feasibility study on tourism and return on their expenditure.

Contributions of community forestry to local infrastructure development

All three case study sites suffer from a lack of infrastructure that might contribute to quality of life and economic opportunity within the local community, including the availability of basic services and facilities. As the lowest ranked administrative units in Nepal, the government's Village Development Committees (VDCs) were the local-level government[1] with responsibility for delivering development activities at the local level. However, ongoing political instability (as discussed in Chapter 2 and in this section) and limited financial capacity has meant that these government bodies have had very minimal effect in delivering local development activities in the study areas. In this context, the majority (85 percent) of interview respondents said that community forestry is making a positive contribution to local community development. As explained by one elite male respondent:

> We do not get financial support from the government for development projects. Our CFUG carries out many community development activities, such as building a primary school for our children, building a road to link our village with the local market, building a saw mill, a rice mill, and candy-making facility, thereby providing employment for local people. We have also established a savings and credit cooperative. The CFUG's fund has been used also for . . . construction of a rural electricity transmission line . . . the CFUG fund is being used so that a drinking water facility can be implemented for members. So CF is doing a lot more than what the government is able to do for local community development activities.

So, we can see that CFUGs have to some extent become 'proxy' local governments. Given the breakdown of local government institutions during the Maoist insurgency (1996–2006), and thus lack of coordination and delivery of services by government into many rural areas, CFUGs have been forced into provision of development activities in their communities. As noted by Rechlin et al. (2007), during this insurgency, VDCs were effectively 'defunct' in some locations, and Nightingale and Sharma (2014) have reported that more than two-thirds of VDC secretaries fled their local communities for the safety of district headquarters. Given that CFUGs were seen by the Maoists to maintain local legitimacy based on being locally based, democratic, community owned rather than government driven, just, equitable and relatively transparent in their processes to benefit all members (Nightingale and Sharma 2014; Rechlin et al. 2007), CFUGs continued their forest protection and management activities, and in the absence of effective government institutions, assumed the role of local government to deliver local services. Enabled by the community forest guidelines which encourage the expenditure of funds on community development activities, CFUGs have continued to play this role even after the conclusion of the insurgency. So, CFUGs, as self-organized entities, are not only actively involved in community forest management, but are investing their funds and voluntary efforts also in local development projects. In the case study areas, the availability of timber at a lower price encourages the building of schools, health posts and members' houses. Other development activities, for which CFUG funds have been spent, with the collaboration of Village Development Committees, include schools, health posts,

Table 5.2 Investments of CFUG funds for infrastructure development (from handover of community forests up to 2012, in NRs)

CFUG expenditures	KCFUG	HJCFUG	SDCFUG	Total of all CFUGs	Percent of total expenditure
Drinking water	1,065,033	64,308	24,188	1,153,529	**2.4**
Rural road	976,562	946,045	83,439	2,006,046	**4.1**
Schools	1,964,243	2,037,704	225,669	4,227,616	**8.6**
Vegetable collection centre	0	6,050	0	6,050	**0.01**
Soil conservation and flood control	151,114	121,000	0	272,114	**0.6**
Temple		8,053	29,231	37,284	**0.1**
Health post support	75,026	0	0	75,026	**0.2**
FUG land purchasing	269,376	0	0	269,376	**0.6**
FUG office building and compound	2,573,321	0	743,680	3,317,001	**6.8**
Culvert/bridge	246,795	0	0	246,795	**0.5**
Total	**7,321,470**	**3,255,760**	**1,172,757**	**11,749,987**	**24**

(Source: data compilation and calculation from CFUG's office records, annual progress reports and audit reports)

drinking water facilities, electricity facilities, irrigation, temples, vegetable collection centres, community buildings and halls, soil and water conservation, and small bridges and culverts (Table 5.2).

Data shown in Table 5.2 indicate that the CFUGs are acting as local development agents. Apart from CFUGs, there are no formal bodies in local communities whereby the protection and management of forest resources can be organized or community funds be allocated for community development activities. Nor are there any other bodies that formally connect the local villages to the VDCs or the next higher formal level of government, the District Development Committees (DDCs).

Indeed, community development projects would not be possible if CFUG funds were not available. In the past, local people used to visit government bodies and request money so that their annual community budgets could be taken care of because communities had no money. However, CFUGs have provided a locally based institutional structure for collection and investment of funds, and have subsequently decided to invest their own funds and voluntary labour in relation to local development activities. So, community forestry is virtually the only way that a wide range of local community developments can occur.

However, local people have different opinions regarding community development activities. For example, most of the local rich and medium-income members hope that the local rural road networks will improve access to and from the villages. Poor people, however, do not regard investment of CFUG funds in the construction of rural roads as important. They believe that the road-building project will not provide any benefit for poor people or their livelihoods. Indeed, due to the construction of the road, poor people are likely to lose their jobs as porters carrying loads of items for wealthy people; hence a perception amongst some poor people that CFUG funds are being used in ways that further undermines their livelihoods.

Community participation in community forestry

Active community participation in CF is a core principle outlined in Nepal's national forest policy (GoN 2015), and is one of the main prerequisites for sustainable community forestry and better socio-economic outcomes for local community development (Pokharel 2010; Devkota 2012; GoN 2015). A sense of forest ownership by local communities and use rights to forests are crucial in order that there be active participation in forest resource management by local communities (Ellsworth and White 2004; Gilmour, O'Brien and Nurse 2005). Moreover, people participate when they believe not only that they have control of the security of forest resources, but will receive all of the benefits that flow from that control (Devkota 2012). So community participation is expected in relation to the overall planning and decision-making processes to deliver benefit sharing based on equity. So, active participation by the local community is indispensable for sustainable community forestry and community development.

Extent of participation on Executive Committees: gender, caste and wealth

In recent years, the increased participation of women and Dalits on CFUG committees indicates that there have some been improvements, in terms of social outcomes, because of CF. When speaking about community participation in forestry, one government forestry officer involved in the establishment and development of the CF program in the Sindhu Palchok District for over two decades said:

> My field observations since the 1980s show that people participate in the community forestry program when they believe that their share of benefits from local forest resources, as a result of equitable participation on the CFUG executive committees, is secure. Equitable participation by women and poor members of the CFUGs on executive committees is gradually increasing. This is a prerequisite for the sustainable management and development of forest resources, regarding support for rural livelihoods.

The CFUGs' records indicate that the Executive Committees (CFUG ECs), which have been formed since the first year of the handover of the forest, have been dominated by high caste wealthy men. Number of households, population and ethnicity of CFUGs are relatively different across these three cases. The populations and number of households of KCFUG are much greater, as is the number of members on the EC in 2012. This is due to migration of people from the Mountain and Middle Hill regions to Chitwan (in the Terai for employment and better life). Table 5.3 shows the total composition of the Executive Committees by caste and gender, up to 2012, from the formation and registration of the CFUGs, which for KCFUG was 1995 and for HJCFUG and SDCFUG was 1998 (see Table 5.3a, Table 5.3b and Table 5.3c). Upper caste men are clearly dominant, making up

Table 5.3a Composition of the KCFUG EC (by caste and gender)

Period	Upper caste			Middle caste			Lower caste			Total		Grand total
	M	F	Total	M	F	Total	M	F	Total	M	F	
1995–1996	12	2	14	5	2	7	0	0	0	17	4	21
1996–1999	12	2	14	1	2	3	0	0	0	13	4	17
1999–2002	10	2	12	3	2	5	0	0	0	13	4	17
2002–2005	9	2	11	4	2	6	0	0	0	13	4	17
2005–2007	10	3	13	2	2	4	0	0	0	12	5	17
2007–2012	10	2	12	0	4	4	1	0	1	11	6	17
Total	63	13	76	15	14	29	1	0	1	79	27	106

Note: M – male and F – female

(Source: KCFUG FOPs, 1995, 2001 and 2006; meeting minutes, 2012)

Table 5.3b Composition of the HJCFUG EC (by caste and gender)

Period	Upper caste			Middle caste			Lower caste			Total		Grand total
	M	F	Total	M	F	Total	M	F	Total	M	F	
1998–2001	11	0	11	0	0	0	0	0	0	11	0	11
2002–2006	11	0	11	0	0	0	0	0	0	11	0	11
2007–2012	6	4	10	0	0	0	1	0	1	7	4	11
Total	28	4	32	0	0	0	1	0	1	29	4	33

Note: M – male and F – female

(Source: HJCFUG forest operational plans, 1998, 2002 and 2007; fieldwork, 2012)

Table 5.3c Composition of the SDCFUG EC (by caste and gender)

Period	Upper caste			Middle caste			Lower caste			Total		Grand total
	M	F	Total	M	F	Total	M	F	Total	M	F	
1998–2000	3	1	4	6	3	9	0	0	0	9	4	13
2001–2002	4	2	6	5	2	7	0	0	0	9	4	13
2003–2004	2	2	4	7	2	9	0	0	0	9	4	13
2005–2006	3	1	4	6	2	8	1	0	1	10	3	13
2007–2008	4	1	5	3	4	7	1	0	1	8	5	13
2009–2012	3	1	4	3	3	6	1	0	1	7	4	11
Total	19	8	27	30	16	46	3	0	3	52	24	76

Note: M – male and F – female

(Source: SDCFUG forest operational plans, 1998, 2001 and 2006; fieldwork, 2012)

51 percent of all CFUG EC appointments, though in SDCFUG, only 25 percent of EC members have been upper caste men (compared to 59 percent in KCFUG and 85 percent in HJCFUG). So, the different CFUGs have clearly offered different opportunities for participation by marginalized groups such as women and Dalit.

Table 5.3d Total membership of CFUG ECs by caste and gender from forest handover up to 2012

CFUGs	Upper caste			Middle caste			Lower caste			Total		Grand total
	M	F	Total	M	F	Total	M	F	Total	M	F	
KCFUG	63	13	76	23	15	38	1	0	1	79	28	107
HJCFUG	28	4	32	0	0	0	1	0	1	29	4	33
SDCFUG	19	8	27	30	16	46	3	0	3	52	24	76
Total	**110**	**25**	**135**	**53**	**31**	**84**	**5**	**0**	**5**	**160**	**56**	**216**

Note: M – male and F – Female

(Source: CFUG FOPs, meeting minutes, 2012)

A review of the minutes of meetings and of other official records of the study group CFUGs shows that the representation, on Executive Committees, by women and members of lower castes has only marginally increased or changed following the handover of community forests. Although involvement by women and lower caste members, on all three Executive Committees of the study group CFUGs, has marginally improved, the data indicate that there is still disproportionate representation of upper castes on the committees (Table 5.3). The general members of the CFUGs are still ignorant regarding the guidelines and compliance provisions in relation to the inclusion of women and poor people on those committees. This situation creates obstacles for enhancing existing capacity of poor, Dalits and women members of CFUGs, and ultimately community development activities have been affected adversely at the local level.

Membership of CFUG ECs is still heavily dominated by people in the medium and rich well-being categories, as shown in Table 5.4. Most of the elites and rich people, who dominate the Executive Committee membership, know the provisions regarding the inclusion of women and poor members, but are reluctant to implement them. In addition, heavy time commitments for household work and general labour cause the majority of women and poor members of the CFUGs to be excluded from participating in meetings and the decision-making process.

The elite members also remain reluctant to share power and are not interested in providing opportunities for women and poor members of their CFUGs to participate on committees. One of the lower caste women of the HJCFUG confirmed that women and poor members are being excluded:

> All is still not going as well as one may think. Women, poor people, and lower caste members, in particular, are still excluded in relation to decision making, as well as the benefit sharing process regarding community forestry. We are only invited by CFUG EC members to meetings when foreigners and new visitors are going to be present; otherwise we are completely excluded.

Although women are involved in CFUGs as forest users with rights to use forest resources, their participation in both general assembly of CFUGs and on Executive Committees is limited. One of the Dalit women members of the KCFUG said that the:

Table 5.4 Total membership of the CFUG ECs by wealth categories from forest handover up to 2012

Period	Very poor	Poor	Medium	Rich	Total
KCFUG	0	1	44	62	107
HJCFUG	0	1	12	20	33
SDCFUG	0	3	27	46	76
Total	0	5	83	128	216

(Source: CFUG's forest operational plans; fieldwork)

lack of female participation on the committee, or decision-making body, not only denies our needs and interests, but it also deprives us of our rights regarding community forestry. . . . Major constraints, regarding our participating in decision making and the benefit sharing process, include lack of availability of time, the increased household and farm work burden, and lack of opportunities and supportive interventions.

It has been observed that wealthy and elite male members of the Executive Committees in all three CFUGs are not in favour of women being included in the decision-making process. As explained by an elite male member of the SDCFUG EC:

Women are best at carrying out household chores that men cannot do. Women do not know about the existing policies and provisions regarding our FOP. So, they just sit and listen. Men can do these things better. We have kept the women's names in the samiti (CFUG EC) because we are required by law to do so.

Indeed, women, poor people and Dalits are the major forest users. So, they have important roles to play in relation to the sustained management of forest resources. However, because of social attitudes (e.g. women should not go to community meetings; women should not speak in front of men) and low literacy levels, they are largely confined to carrying out household or similar chores (Devkota 2012). Any change apparent in EC membership over time indicating a greater level of participation by marginalized CFUG members may in fact be misrepresenting the actual power relations and operations of these committees for meaningful participation in community development. So, in terms of community development, women and Dalits still have limited participation in decision-making, thereby restricting their opportunities to enhance their livelihoods.

Community participation in protection and management of community forests

Active community participation in protection, management and utilization of products from CF is one of the most important elements for sustainable management and development of forest resources, subsequently enabling CF to be a process for community development. In the past, one of the causes of deforestation and degradation of government-managed forests was the lack of participation by people and the opportunistic use of forest resources. One senior forestry professional of the Ministry of Forests and Soil Conservation explained how important local participation is and the reasons that people should partici-pate in forest protection and management:

> Community forestry is much better. In the past, the government was not able to properly protect forest resources. Before the commencement of com-munity forestry . . . the government was not able to convince people of the significance of forest management, regarding their day to day livelihoods. Local people were not really obtaining any benefit from government-man-aged forestry. That's why they were not participating in the protection and management of the forests.

The household survey, focus-group discussions and in-depth interviews have revealed that CFUGs are protecting their forests by making sure that there is no unauthorized tree felling, poaching of wildlife, shifting cultivation, encroach-ment, overexploitation of forest products and unregulated grazing (if it is permit-ted at all). About 91 percent of respondents to the household survey (n=138) believed that there has been an improvement in the level of participation of women and poor people in relation to the protection and management of com-munity forests (Table 5.5).

Respondents in all three CFUGs have said that the major cause of fire before the handover of the forests was the lighting of fires by forest users to encourage fresh shoots of grass. Sometimes fires were lit by children who were in the forests playing. However, the number of forest fires has reduced since the formation and establishment of the CFUGs. The villagers gave up setting fires when the CFUGs decided to enforce rules regarding the protection of their forests. In relation to all three CFUGs, there are provisions in their FOPs for the protection of com-munity forest land. Respondents have indicated that there are three main meth-ods whereby forest can be protected. These are: (a) the use of employed forest watchers, (b) patrolling by users and (c) everyone following the rules regarding protection.

Table 5.5 Perception of CFUG members regarding participation of women and poor in forest protection and management (n=138)

CFUG	Do not know (%)	Improving (%)	Not improving (%)
KCFUG	7	92	1
HJCFUG	9	89	2
SDCFUG	8	92	0
Overall	8	91	1

(Source: household survey, 2009)

Capacity building as a means of community development through community forestry

An increase in capacity of CFUGs for community development including through improved literacy status, enhanced skills and knowledge, empowerment of women and Dalits, and leadership development has been reported in the three case study sites since the CF program commenced.

The capacity of CFUGs to develop local leadership and management skills, liaise with government staff, and empower women and poor people is an important social outcome flowing from people's involvement in CF processes. The case studies provide evidence that CF can be effective not only in improving the management of forest resources, but also in transforming the social conditions within local communities, as described throughout this section.

Changes in literacy status

The results of the household survey indicate that the literacy levels in all CFUGs have increased as a result of CFUGs spending funds on adult literacy programs and local school support programs. All three CFUGs have shown a substantial increase in population literacy rate from the handover of forests and establishment of CFUGs up to 2009 (a period of 14 years from 1995 for KCFUG and 11 years for HJCFUG and SDCFUG). In 2009, the literacy rate was highest in KCFUG at 72 percent, an increase from 56 percent in 1995. In HJCFUG, literacy increased from 52 percent in 1998 to 70 percent in 2009, and SDCFUG has shown an increase from 53 percent in 1998 to 69 percent in 2009 (see Figure 5.2).

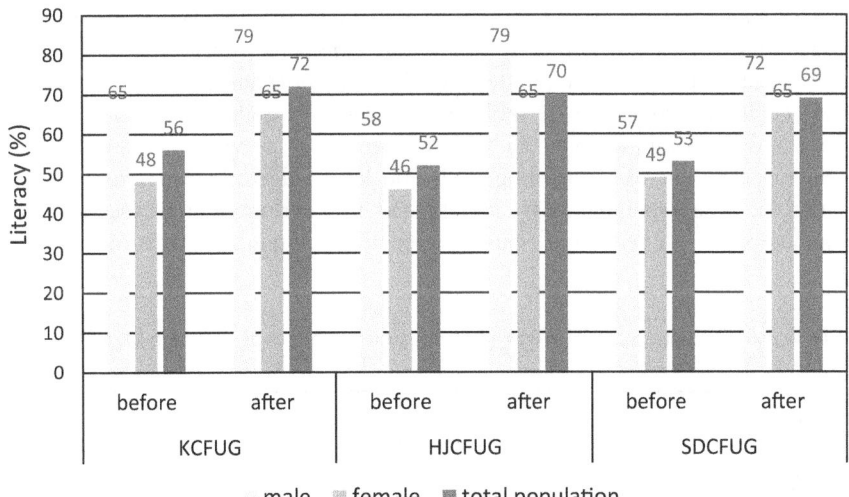

Figure 5.2 Change in literacy status following the implementation of the community forestry program

(Source: household survey, 2009)

Both men and women have participated in the literacy programs, resulting in increases in both male and female literacy rates. Before the introduction of the community forestry program, literacy rates amongst women were below 50 percent across all three CFUGs, whereas by 2009, all three CFUGs had achieved a rate for women of 65 percent. Interestingly, while the differential between the male and female rates has fallen in KCFUG and SDCFUG, the differential in HJCFUG has actually increased. One female respondent, a participant in literacy classes organized by KCFUG, commented:

> The main reasons that the literacy of women has increased is the literacy class run by the CFUG; and that more girls are attending nearby schools because of improvements in the local school facilities because of investment by the CFUG.

As well as running adult literacy programs, another reason that the literacy rate has gradually increased in the villages is that the CFUGs have invested funds in school education. New primary schools have been established and former primary schools are being upgraded to secondary schools. The utilization of CFUG funds to support local education has been a major development. However, in all CFUGs, male literacy rates remain higher than females.

Changes in skills and knowledge

Following the implementation of the community forestry program, the CFUGs have received training to strengthen existing skills and knowledge to support income generation. The awareness level of users in relation to conserving biodiversity has increased, as has their involvement in various income generating activities (such as sericulture, goat farming, and micro-enterprise development).

About 72 percent of respondents of the household survey said that members are gradually acquiring new skills and knowledge through the implementation of the CF program (Table 5.6). Only 28 percent of respondents said that there has been no increase regarding skills and knowledge or did not recognize an increase. The respondents have asked for additional training in various activities, including record keeping, document preparation, financial management, forest

Table 5.6 Perception of CFUG members regarding changes in skills and knowledge from CF (n=138)

CFUG	Do not know (%)	Improving (%)	Not improving (%)
KCFUG	15	65	20
HJCFUG	20	70	10
SDCFUG	10	80	10
Overall	**15**	**72**	**13**

(Source: household survey, 2009)

management, collection procedures and the efficient use of CFUG funds. The respondents have also indicated that any training and extension programs should be implemented with proper monitoring and evaluation.

Not all respondents are satisfied with progress regarding the acquisition of new technical skills and knowledge, expressing a desire to obtain further skills and knowledge through more training and extension programs. Some respondents have suggested that special training and workshop programs be set up for women and poor members, especially in remote areas. As explained by the chairperson of the Shreechhap Deurali CFUG:

> We (CFUG EC members) are relatively better off, in having acquired some skills and knowledge. But the majority of members are still technically weak and somewhat backward. We just do not have enough resources and skilled people in order that socio-economic conditions improve.

Despite some limitations in training reported by CFUG EC members during interviews, the implementation of CF programs has improved the skills and knowledge of some people in the CFUGs.

Social network formation

The formation and development of social networks as social capital play an important role in community development. Social capital can be seen as the attributes of a community or group (such as networks, norms and trust relationships) that facilitate and promote collective action for mutual benefit (Ojha 2006; Putnam, Leonardi and Nanetti. 1993). As a result of community forestry strengthening social networks at the local level, new community development opportunities have emerged for planning and implementation of community development activities. While not looking in detail at the concept of social capital, for the purposes of this chapter, social capital includes establishing intra- and inter-community linkages through social networks, whereby positive relationships form and social cohesion is enhanced.

The case study research found that the CF program has contributed to the establishment of a system of networks and trust relationships which enable communities to address common problems. The potential benefits of enhanced social capital include community involvement in forest management and development, income attainment, the reduction of rural poverty and community development. Social networks help create a conducive environment for social capital formation. So, local CFUG members may experience the benefits of increased social capital in a number of ways, including through links beyond the CFUG with other stakeholders and development partners who provide technical and financial support for planning and implementation of community development activities, and through greater social cohesion through which CFUG members can effectively interact in the planning and implementation of community development activities at the local level (Devkota 2012).

Social networks of CFUGs

The results of the case studies indicate that each of the three CFUGs and their communities have developed external social networks through the community forestry program. Community forestry acts like an umbrella organization or conduit, in that it provides various social services in rural areas. One DOF senior forestry official said that:

> By comparison to the past, community forestry is playing a key role regarding increasing linkages between CFUGs and various governmental and non-governmental institutions.

All three CFUGs have established networks with various government and non-government offices in order that services be provided for their members (Figure 5.3). For example, all CFUGs retain a close working relationship with the District Forest Office (DFO) and the *Ilaka* (Area Forest Office), which provide legal and technical assistance in support of forest management and development. Other government offices, such as the District Agriculture Office and the Agriculture Service Centre provide the CFUGs with technical advice to assist in cultivation of a variety of fruits, vegetables and other crops; as well as controlling pests and preventing diseases. Similarly, technical advice and assistance from the District Livestock Development Office and Livestock Service Centre support CFUGs not only in production of new varieties of fodder and new breeds of livestock, but also in the effective treatment of sick livestock. Village Development Committees also collaborate with CFUGs in development of local community infrastructure. Most of the local NGOs and government agencies which

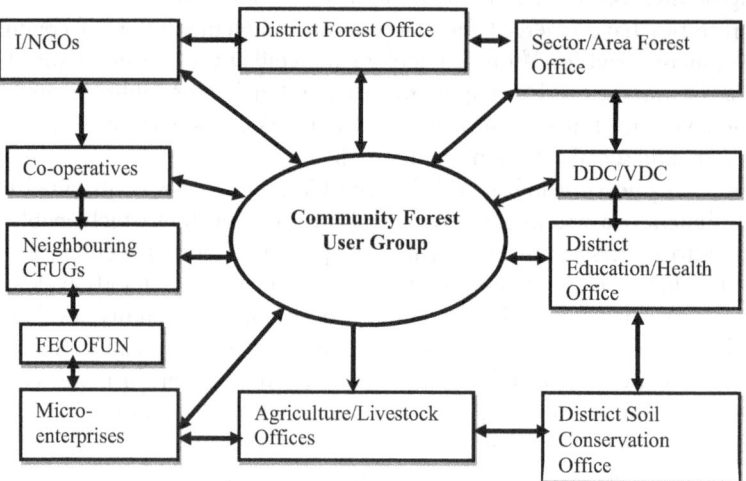

Figure 5.3 Example of a typical social network of CFUG

(Source: field research 2009, 2011)

implement programs with community participation have established networks with CFUGs. Social networks are important, especially in terms of help being provided regarding the lending of money or the supply of goods and services (e.g. mobile health-camp, rescue and relief operations during natural calamities); sharing of skills, knowledge and experiences; marketing of goods and services; and so on. Thus, social networks provide platforms for integrated local community development programs.

Many of the key informants (n=138) reported that the benefits that flow as a result of stronger social networks are that:

- new information is provided about resource management,
- new opportunities arise in relation to obtaining benefits from different agencies and from new expanded markets; and
- confidence and trust amongst partners increases.

CFUGs develop their own networks through co-operatives and through other local institutions. The case study CFUGs have gone beyond forest protection and management, developing important roles in community development planning and implementation. They do this through various networks and linkages so that conflicts can be resolved, ideas and experiences can be shared, resources and techniques can be made available, and planned activities can receive support. As reported by 93 percent of the respondents of the CFUG ECs, networks have been emerging through implementation of community forestry (Table 5.7).

The FECOFUN is assisting some of its members in relation to the preparation and implementation of good-quality constitutions and FOPs, in order that these can become models for CFUGs in other districts. CFUG members are becoming increasingly active in relation to networking and they have established very good networks with local government and non-government organizations.

Social cohesion within CFUGs

Social cohesion refers to trust between members within a CFUG, with evidence that social cohesion can be developed through the CF group process. During focus-group discussions, the respondents reported improved social cohesion and trustworthiness, describing a range of indicators of the contribution of CFUGs to social cohesion (Box 5.1).

Table 5.7 Perception of CFUG members regarding CFUG networks (n=138)

CFUG	Do not know (%)	Improving (%)	Not improving (%)
KCFUG	8	90	2
HJCFUG	6	94	0
SDCFUG	2	95	3
Overall	**5**	**93**	**2**

(Source: household survey, 2009)

Box 5.1   Some indications of improved social cohesion for community development in the CFUGs

- Help and support have been provided in emergencies
- Members of the groups are perceived to be closer than they were before the implementation of CF
- Cooperation between members has increased
- Mutual trust between different members has increased
- The gap between various groups of people (in terms of sex, caste/ethnicity, income, religion, origin, education, political party) has narrowed
- People have learned from each other
- Contacts with external agencies have increased
- Transparency, regarding the use of forest products and funds, has increased, and consequently there has been a reduction in corruption regarding community forestry

(Source: focus-group discussions, 2009)

Research participants in interviews and focus-group discussions from all three case study CFUGs described an increase in social cohesion arising from the implementation of CF, particularly given that community related problems and issues can be discussed during CFUG meetings. People have spoken about CFUG meetings as being the time to discuss local problems. Thus, they believe that the CF program has helped not only to reduce conflicts between forest users (for example conflicts regarding the utilization of forest resources), but also reduce conflicts between the local community, the DFO and other government authorities.

Discussion

Some of the major areas regarding the contributions of CF to community development are discussed in the following section.

Community forestry provides a promising platform for local leadership

The existing CF guidelines contain certain provisions whereby affirmative action can take place. They also contain provisions that look after the interests of poor people, women and occupants of marginalized households. Furthermore, the guidelines contain a provision regarding proportionate representation, on decision-making bodies, of women, different castes and ethnic groups (DoF 2014). There have been some indications that women and poor people are becoming more active members of their local CFUGs and therefore becoming more active actors in broader community development processes with improved livelihoods.

Before the implementation of the CF program across Nepal, some disadvantaged people could not speak confidently in public, or even share their views and ideas with elite or male members of their communities, because of their lower social status and fear of reprisals. After the implementation of the CF program, and the engagement of disadvantaged people in the community forestry process, some of those disadvantaged people have become empowered to some extent (Devkota 2012). They have benefited through various community forestry training programs and workshops, in such a way that they have learned to express their views and share ideas with other members of their communities, and can even speak with outsiders. Participation of women, poor and disadvantaged groups in community development activities has seen a gradual, though limited, increase with improved leadership skills and empowerment. Small steps have been taken, but there is still more change required.

Community forestry enables establishment of networks that contribute to community development

Community forestry in Nepal has led to the development of decentralized social networks based around the CF institutions and the associated communities themselves. So, CF in Nepal can be regarded as being an important form of decentralized forest management. CF has led to the creation of robust institutions throughout Nepal over a number of years, which are now regarded as having great importance in the implementation of community development activities and improvement of the livelihoods of rural people (Pokharel 2008). Community forestry plays a key role in increasing the links between CFUGs and different governmental and non-governmental institutions (Devkota 2012) for planning, implementation, monitoring and evaluation of community development activities. CFUGs have even established links at the international level. FECOFUN is an example of a CFUG network which has strong links at the local, national and international levels, whereby government can be urged to continue implementing a forestry policy that provides positive outcomes for local community development.

CFUGs have the potential to act as local service providers

Over more than 30 years, CF in Nepal has evolved from being a small exploratory forest management program to becoming an important and widespread national movement, involving more than 35 percent of the country's population (DoF 2017). The emphasis is on making sure that the supply of essential forest products is sustainably managed and that rural community development occurs, given that CFUGs can benefit from access to key institutional and technical services. The emergence of CFUGs as Community-Based Organization (CBO) service providers has been another achievement from community forestry in Nepal. Facilitators from capable CFUGs provide services to less capable CFUGs, thus delivering a farmer-to-farmer extension approach resulting in learning being disseminated in a cost-effective way. CFUGs have also displayed some resilience in the face of local conflict and instability, maintaining their institutional structures and functions while local government institutions have collapsed or become ineffective, and thus have been able to continue to support their local communities in the absence of government services and support.

Community forestry as a vehicle for local community development

Improved access to forest resources can help increase the availability of forest products for construction purposes and for other economic benefits (Pokharel 2009). The community forestry fund can be used to assist poor and disadvantaged members of CFUGs (Baral and Stern 2011; Nath and Inoue 2010; Pokharel 2009). Case study findings indicate that CF could be seen as a rural community development program, in which virtually all aspects of development can be implemented through community participation and the investment of CFUG funds. Over the last 15 years, CFUGs have substantially contributed to a wide range of community services, such as infrastructure development and forest resources conservation. CFUGs provide a forum for local community members to discuss local development issues and share their experiences and ideas with other users. CFUGs also provide a contact point for development agencies. So, CFUGs can be useful institutions for planning and delivery of local community development interventions.

About 24 percent of the total income of case study CFUGs is invested in local community development, such as the construction of schools, community buildings, roads, culverts, and drinking water facilities (Devkota 2012). The decision-making process regarding investment of CFUG money in local community development activities is still largely controlled by elite members of the community. So investments in specific pro-poor activities or facilities are not usually priorities of CFUGs. Mostly, it is the rich and medium-income CFUG members who are in the best position to take advantage of most community development activities (Devkota 2012; Yadav 2004). For example, while any improvement, regarding school facilities, can be of considerable benefit to rich members, it can be of far less benefit to poor people whose children do not attend school, as they have to work at home to look after animals, collect fuelwood and care for younger siblings.

Furthermore, investment in construction of rural roads helps to link villages with local markets. Users who produce buffalo milk, green vegetables, potatoes and other cash crops on a commercial scale benefit more from the building of such roads than poor users who have virtually no surplus agriculture products for marketing (Devkota 2012). While community development projects delivered by CFUGs may appear impressive, CFUG funds are mostly being invested or spent according to the interests and priorities of elite members of rural communities.

Those interests and priorities include electrification, irrigation and roads that directly benefit the richer people in the communities.

Conclusions

This chapter has shown that CF in Nepal has delivered socio-economic benefits for families engaged in enterprise development activities and broader outcomes for local community development activities (e.g. drinking water facilities, roads, schools) from investments made by CFUGs. Sustainable forest management and development of forest resources is occurring when there is active participation by all members of the CFUG, including women, poor people and other disadvantaged people. In addition, CFUGs have been able to invest income from the sustainable harvesting of forests over the last 15 years in formal school education facilities; informal education programs for women, poor and disadvantaged members; as well as scholarships for students from poor families. CFUG funds have also been utilized in development of local infrastructure and alternative income generating activities.

Since community forestry was implemented in Nepal, social mobilization has gradually increased in rural areas. That mobilization includes (a) community participation in the community forestry process; (b) the development of forest management skills and knowledge amongst CFUG members; and (c) the formation and strengthening of social networks. Apart from social mobilization, the improvement in forest condition arising from implementation of the community forestry program has enhanced both the quantity and quality of forest products available.

Nonetheless, up till now, there has been a limited positive impact on the livelihoods of poor and other disadvantaged CFUG members because the protection-oriented forest management approach of CFUG ECs has not made available the required amounts of forest products to all CFUG members (see Chapter 4). The inequitable distribution of benefits from community forestry arises primarily because of the domination of the distribution process by elite members. Furthermore, the collecting and mobilizing of CFUG funds for community development are still controlled mainly by elite male CFUG EC members. So, contributions by CFUGs to local community development programs still depend largely on the interests and decisions of those elite members of CFUG ECs. Money is spent mainly on the construction of buildings, roads and temples, whereby poor people do not directly benefit.

Gilmour (2016) cautions that even when the pre-conditions for successful CF exist, it may still take many years for the local social and institutional transformation to occur and the ambitions for CF to be achieved. Therefore, in order for the objectives of CF to be reached, policymakers and forest managers have to be able to recognize what is required, in relation to important socio-economic outcomes for community development being achieved, and the full rights and capacity of people to participate in the management and development of forest resources properly established and sustained.

Note

1 At the time this research was completed, VDCs were the local government administrative units; however under the requirements of the new constitution (2015), VDCs were dissolved in March 2017 and replaced by new rural municipalities, or Gaupalika, with increased budget and taxing powers.

References

Adhikari, J.R. and Adhikari, B. 2010, *Conflicts and community forestry: Understanding the impact of the decade-long armed conflicts on environment and livelihood security in rural Nepal*, paper presented at CAPRi Workshop on Collective Action, Property Rights, and Conflict in Natural Resources Management, 28 June–1 July 2010, Siem Reap, Cambodia, viewed 7 May 2017, www.researchgate.net/publication/228919465_Con flicts_and_Community_Forestry_Understanding_the_Impact_of_the_Decade-Long_Armed_Conflicts_on_Environment_and_Livelihood_Security_in_Rural_Nepal

Baral, N. and Stern, M.J. 2011, A comparative study of two community-based conservation models in Nepal, *Biodiversity Conservation*, vol. 20, no. 11, pp. 2407–2426.

Bhattarai, R.C. 2011, Economic impact of community forestry in Nepal: A case study of mid-hill districts of Nepal, *Economic Journal of Development Issues*, vol. 13 and 14, no. 1–2, pp. 75–96.

Cavaye, J. 2006, *Understanding community development*, Cavaye Community Development, viewed 12 March 2017, www.southwestnrm.org.au/sites/default/files/uploads/ihub/understanding-community-developments.pdf.

Chhetri, R.B. 2006, From protection to poverty reduction: A review of forestry policies and practices in Nepal, *Journal of Forest and Livelihood*, vol. 5, no. 1, pp. 66–77.

DeFilippis, J. and Saegert, S. 2012, Communities develop: The question is how? In J. DeFilippis and S. Saegert (eds), *The Community Development Reader*, 2nd edn, Routledge, New York.

Devkota, B.P. 2012, *Socio-economic outcomes of community forestry in Nepal: Lessons from three diverse rural communities*, PhD Thesis, School of Environmental Sciences, Charles Sturt University, Australia, viewed 25 May 2017, http://primo.unilinc.edu.au/primo_library/libweb/action/dlDisplay.do?vid=CSU2&docId=dtl_csu40107

DoF 2014, *Community Forestry Development Program Guidelines*, 3rd edn, Department of Forests, Government of Nepal, Kathmandu, Nepal.

DoF 2017, *Community forestry*, Department of Forests, Government of Nepal, viewed 23 May, 2017, http://dof.gov.np/dof_community_forest_division/community_forestry_dof

Ellsworth, L. and White, A. 2004, *Deeper Roots: Strengthening Community Tenure Security and Community Livelihoods*, Ford Foundation, New York.

Gilchrist, A. and Taylor, M. 2016, *The Short Guide to Community Development*, 2nd edn, Policy Press, University of Bristol, UK.

Gilmour, D. 2016, *Forty Years of Community-Based Forestry: A Review of Its Extent and Effectiveness*, Food and Agriculture Organization of the United Nations, Rome, Italy.

Gilmour, D.A., O'Brien, N. and Nurse, N. 2005, Overview of regulatory frameworks for community forestry, in N. O'Brien, S. Matthews and M. Nurse (eds), *Regulatory Frameworks for Community Forestry in Asia*, First Regional Community Forestry Forum, Proceedings of a Regional Forum, RECOFTC, Bangkok, Thailand, pp. 3–33.

GoN 1993, *Forest Act, 2049 (1993)*, Government of Nepal, Kathmandu, Nepal.

GoN 2015, *Forest Policy 2015*, Ministry of Forests and Soil Conservation, Government of Nepal, Kathmandu, Nepal.

Mahanty, S., Gronow, J., Nurse, M. and Malla, Y. 2006, Reducing poverty through community based forest management in Asia, *Journal of Forest and Livelihood*, vol. 5, no. 1, pp. 78–89.

Nath, T.K. and Inoue, M. 2010, Impacts of participatory forestry on livelihoods of ethnic people: Experience from Bangladesh, *Society and Natural Resources*, vol. 23, pp. 1093–1107.

Nightingale, A. and Sharma, J. 2014, Conflict resilience among community forestry user groups: Experiences in Nepal, *Disasters*, vol. 38, no. 3, pp. 517–539.

NSCFP 2007, *The Multi-Partnership Approach: NSCFP Experiences of Working with Multiple Partners*, Nepal Swiss Community Forestry Project, Kathmandu, Nepal.

Ojha, H.R. 2006, Techno-bureaucratic doxa and challenges for deliberative governance: The case of community forestry policy and practice in Nepal, *Policy and Society*, vol. 25, no. 2, pp. 131–175.

Pawar, M.S. 2010, *Community Development in Asia and Pacific*, Routledge, New York.

Pokharel, R.K. 2008, *Nepal's community forestry funds: Do they benefit the poor?* SANDEE Working Papers, ISSN 1893–1891; 2008- WP 31, Kathmandu, Nepal.

Pokharel, R.K. 2009, Pro-poor programs financed through Nepal's community forestry funds: Does income matter, *Mountain Research and Development*, vol. 29, no. 1, pp. 67–74, www.bioone.org/doi/pdf/10.1659/mrd.996

Pokharel, R.K. 2010, Development of community infrastructure through community forestry funds: What infrastructure gets priority, *Banko Janakari*, vol. 20, no. 1, pp. 44–50, www.nepjol.info/index.php/BANKO/article/view/3508/3030

Putnam, R.D., Leonardi, R. and Nanetti, R.Y. 1993, *Making Democracy Work*, Princeton University Press, Princeton, NJ.

Rechlin, M.A., Burch, W.R., Hammett, A.L., Subedi, B., Binayee, S. and Sapkota, I. 2007, Lal salam and Hario ban: The effects of the Maoist insurgency on community forestry in Nepal, *Forests, Trees and Livelihoods*, vol. 17, pp. 245–253.

Springate-Baginski, O. and Blaikie, P. 2003, *Is Community Forestry in Contemporary Nepal Pro-Poor and Sustainable? A Policy Process Analysis*, Stockholm Environment Institute, York.

Yadav, N.P. 2004, *Forest user groups in Nepal: Impacts on community forest management and community development*, Unpublished PhD Thesis, The University of Leeds, School of Geography, UK.

6 Forest tenure and community forestry in Nepal

Trends and implications

Ganga Ram Dahal, Krishna Adhikari and Richard Thwaites

Introduction

Generally national governments assert authority and control over land and natural resources and exploit them as a means of generating and augmenting revenues. The idea that the land and other resources are owned by a sovereign monarch (the crown), and, thus by extension, by the state has European origin (Esman and Uphoff 1984). In developing countries which once came under the European colonial powers, the idea of state as the owner and manager of all land and forests was imposed, and this colonial legacy continues to be a basis of forest and land tenure systems even in countries such as Nepal which were never colonized (Clark 2000).

Primarily who manages land and forest resources, under what conditions and for what purposes can be systematically studied using a framework of tenure, which, as we will see, can be defined as the bundle of rights including access, use, management, exclusion and alienation. These sets of allocation or holding of rights pertaining to the management of forest resources determine particular tenure models or regimes, with different sets of outputs and benefit mechanisms. Certain tenure arrangements benefit large investors and governments at the expense of environment and livelihoods of forest-dependent people. Other regimes that encourage users as owners and managers have totally different outcomes (Larson and Dahal 2012; Agrawal and Ostrom 2001). Even within a given regime the types of tenure instruments in place and the degree of tenure security is important in shaping the outcomes (FAO 2011).

Considering the consequences and externalities, forest tenure regimes are changing and more diverse forms of tenure regimes are emerging. For example, the *statist* regime initially adopted by the colonial rulers invited severe consequences to the environment and livelihoods of forest-dependent people (Ribot 2002). To minimize the damage and bring about positive changes, nation states have sought alternative models of governing the forests. In the past few decades, they have emphasized the adoption of progressive measures which seek to strike a balance between public and private ownership and management of forests and forest resources. Today, around 80 percent of total forest land in the world is still under government administration (FAO 2015). In Asia, this picture is slightly different, with 60 percent of the forest controlled and managed by the government, while the remaining 40 percent is distributed for administration by indigenous peoples, local communities or private firms (RRI 2014).

Forest land tenure in Nepal has a long evolutionary history. Earlier chapters in this volume show that Nepal has witnessed a range of forest and natural resource management regimes including 'feudal' tenure until the nationalization of forests in 1957, a traditional customary system often operating informally within the feudal system, and state-owned and -controlled systems since 1957. Over the last four decades, forest tenure in Nepal has undergone a major shift as government has piloted different tenure models in an attempt to ensure sustainable management of resources and to contribute to the livelihoods of local people (Pokharel et al. 2008; Malla 2000). Amongst them are various community-based forest management approaches, including community forestry, collaborative forest management, leasehold forest management and buffer zone community forestry.

Methodological approach

Despite the fact that forest tenure is one of the key factors that influence the achievement of forest management objectives (e.g. improved forest condition (Chapter 3), economic development, improved livelihoods (Chapters 4 and 5), and increased rights of local communities and indigenous groups), studies related to natural resource management in general and community forest resource management in particular often do not consider tenure aspects. In this chapter we aim to look at tenure frameworks and their implications for forestry in Nepal with particular focus on community forestry. By disaggregating the bundle of rights under the tenure systems, we aim (through review of literature and our own observations) to explore and compare what tenure-related factors influence outcomes of different community-based regimes, and to what extent tenure reform has supported (or not) achieving the goals of forests in contributing to people's livelihoods and conservation of biodiversity. We begin this first by defining tenure and associated concepts and looking through the tenure lenses at the evolutionary trend of forest management in Nepal. We then discuss the trend, scales and outcomes of reformed forest policies, particularly community forestry, in different countries (based on a recent study undertaken by FAO) in Asia and how these policies have informed and promoted each other. Finally, the chapter raises questions and issues concerning tenure and the future of community forestry in Nepal.

Concept of tenure in forestry

During the twentieth century, a classic property rights framework has evolved that divides property under three broad categories: public, common and private with state, collective and individual holders of rights, respectively (Larson, Barry and Dahal 2010). Tenure regulates access to and use of resources (FAO 2011; Fisher 1989), but over time, the term 'tenure' has popularly come to refer to various arrangements that assign rights to, and conditions on, those who hold land, ignoring the resources attached to that land. At present, tenure is considered as a contested concept and is understood differently by different actors. In many instances tenure is understood as a narrow concept, which refers to land use rights, and often excludes tree or forest resources. In order to avoid confusion, in our discussion of the concept, we define tenure in a broader sense, incorporating rights to land, but also to the resources associated with that land.

Tenure can be *de jure* or formal (defined by legal statute) and *de facto* or informal (e.g. according to customary practices) (Adhikari and Lovett 2006). Tenure entails a relationship between people as individuals or groups with respect to land and resources and can be either legally or customarily defined (FAO 2011; Larson, Barry and Dahal 2010). As we take it and as it is applied to forest governance, forest tenure is a broad concept that shapes the relationship between people with respect to forests by defining who can use what resources, for how long and under what conditions. So tenure defines the relationship between people and resources with regard to rights and duties.

White and Martin (2002) provide a helpful analytical tool of tenure, defining it as a bundle of rights which encompasses rights to access, use, manage, exclude and alienate. RRI (2012, 2014) has added two more rights to the bundle such as duration of use or ownership and extinguishability. These rights, taken together, can broadly be divided into three parts: operational, collective choice and security related. Operational rights include access rights (the right to enter a defined physical space) and withdrawal or use rights (the right to obtain benefit from a resource such as through harvesting fuelwood, timber or other forest products). Collective choice rights include the rights over management, exclusion and alienation. Management rights include "the right to regulate internal user patterns and transform the resource" (Agrawal and Ostrom 2001, p. 489); for example making decisions over forest management such as silvicultural practices. Exclusion rights entail the right to determine who will have access to the forests and who is excluded. Alienation rights allow selling or leasing the ownership, management and exclusion rights or to use them as collateral. Security-related rights are related to duration and rights of extinguishability, where duration indicates whether the rights are time bound (e.g. as leases) or granted in perpetuity, and extinguishability or rights to compensation indicates whether due process and fair compensation are legally protected if rights are revoked or extinguished.

'Ownership' is an important element in the bundle of rights. It refers to a particular type of tenure in which a more complete bundle of rights is allocated to the rights holders. It implies more or less exclusive and permanent rights including the right to sell or alienate resources. White and Martin (2002) have divided ownership over forest and forest land broadly into two categories: public and private. Generally, public ownership either refers to land and resources owned and administered by the government, or a regime designed and designated for use and management by community, private entity or indigenous people without handover of full 'ownership' rights (RRI 2014). Under private ownership, land and resources are owned by individuals, a community or indigenous people or individual firms. This is a simplified categorization. In reality, land and resource rights arrangements are often more complex, particularly when it comes to customary tenure (RRI 2014).

Since the mid-1980s many countries have initiated rapid changes in resource tenure pattern by introducing decentralization and involving community and private sector in forest management (Gilmour, Durst and Shono 2007). Forest/ land tenure reform entails a progressive change in the rights available to local communities for access and use of forest resources. One of the key features of reform is that the rights may be devolved from the state to local actors (which may include local government), particularly local communities ensuring utmost possible security of such rights.

Studies have shown that in the situation of tenure reform, providing more rights to the community can help achieve effective resource management (RRI 2012; FAO 2006). This is because strong provision of rights provides security and boosts confidence, which in turn foster an environment for the investment of time and other resources by key stakeholders into forest management. Security of tenure expressed through robustness of the instruments assures stakeholders of their ownership for productive investments for sustainable resource use (FAO 2006). Therefore, ownership, duration, robustness, assurance and exclusivity are considered as legal components of secure tenure arrangements (FAO 2006).

Evolutionary history of forest tenure policies in Nepal

During the feudal Rana period (1846–1951), the tenure regime in Nepal was primarily influenced by the need to generate revenue for the state. In the subsequent periods the conservation of forests and subsistence needs of the local people have become the prominent agenda driving forest tenure reform. From the tenure point of view, evolution of forest tenure policies in Nepal can be discussed under three phases: pre-1951 as feudal forestry, the period between 1951 and 1970 which saw collapse of the Rana regime and nationalization of forestry (in 1957), and 1970 onward as decentralization of forestry. There is a common belief that the nationalization of forests in 1957 led to increased deforestation, although there is an alternative view that the deforestation that resulted was largely the result of the political chaos in the 1950s and 1960s. (See Chapter 2 for discussion of this issue.)

Before 1950 the Shah kings and the Ranas ruled the country, often without any constitution or any formal rules even though social organization was maintained by codifying rules based on caste hierarchy (Hofer 1979). Ruling elites held powers to control the forests and other resources and could give away land and forest to their servants and local functionaries either as reward for services or to augment their revenue base, using various tenure instruments, such as *Raikar* (tenure that privatizes land to generate tax revenue), *Birta* (tenure that provides land temporarily for livelihood), *Jagir* (tenure that provides land in return for government service) and *Kipat* (a form of communal land tenure granted to specific groups) (Regmi 1963). Initially, the forest was abundant and was mainly managed in customary ways by indigenous people and local communities. However, with the allocation of tenure rights to those associated with the ruling elite for exploitation and revenue generation, local people dependent on the forest were adversely affected. With a view to increase collection of revenue, (labour) migration from India was encouraged and concessions were granted to landlords to clear the thick forest and convert it to agriculture land (Regmi 1977; Whelpton 2005). Since the feudal basis of forest administration of this period completely ignored the rights and livelihood needs of millions of forest-dependent people, the forest started to decline and was degraded due to poor management and over-harvesting.

Between the collapse of the Rana regime in 1951 and 1970, new policies were formulated governing the forests in Nepal. Through the key regulatory frameworks (Private Forest Nationalization Act 1957; Forest Act 1961; and Forest Protection Special Act 1967), the government nationalized the private (feudal) forests, which according to Regmi (1977) was intended as a progressive reform contrary to elite control of the forests. However, the move to government control also weakened customary practice which had often been tolerated by the feudal owners. Thus, top-down forest policies effectively restricted the legal access to and use of forest products by forest-dependent people, which resulted in detrimental effects on local livelihoods. As a consequence, forest-dependent people, who included marginalized and landless people, often resorted to 'illegal' means in the extraction of forest products.

In response to the alarming rate of forest degradation, there was some realization in the 1970s of the need for people's participation with some rights and ownership, rather than treating them as a destructive force to be policed and managed. This period also saw shifts in the international development discourse giving rise to the idea of decentralization (Adhikari 2015). This impacted forest tenure systems in Nepal. As a first step towards decentralization, regulations were issued in 1978 which created Panchayat Forests and Panchayat Protected Forests to assign forest protection and management rights and responsibilities to the elected local bodies. However, this mechanism was simply a variation of management by the state as the local bodies became the mere instruments of policing the forest-dependent people. Rather than ensuring any rights to the local users and communities, it fuelled conflict between traditional forest users and local bodies (Yasmi, Kelley and Enters 2010).

It is important to note that many local or 'indigenous' systems of forest management persisted or were initiated in the decades following the nationalization of forests. These local systems had no legal status, but often operated in parallel with the ineffective national system of forest management. They were often initiated in response to a locally perceived vacuum in effective forest management (see Chapter 2).

Community tenure in forestry and its outcomes

After a decade of chaotic implementation and limited uptake of panchayat forestry, a major policy shift governing forest tenure was made through the formulation of the Forestry Sector Master Plan in 1988. For the first time, this plan envisioned the involvement of communities in the management of forest resources through the transfer of certain rights. This was later institutionalized through regulatory frameworks including the enactment of the Forest Act 1993, Forest Regulations 1995, and supporting directives and guidance. The Forest Act 1993 has been widely considered as being progressive legislation and a key milestone in establishing community-based tenure in Nepal. The changed policies essentially focused on community and leasehold forestry and redefined the purpose of national forests. The most important community-based tenure regimes in forestry in Nepal include: community forestry, leasehold forestry, collaborative forestry, and buffer zone community forestry. In all these public tenure community-based management regimes, the transfer of rights from the state to the community is primarily for management and use only; ownership of the forest is retained by government. In these regimes many management decisions are taken by local user groups, but the scope of decisions is often constrained by administrative regulations and decisions by the District Forest Office.

Community forestry and leasehold forestry were formally started as national programs in 1993, although there had been significant numbers of pioneer efforts since the late 1980s, especially, but not only, involving the Nepal-Australia Forestry project in Sindhu Palchok and Kabre Palanchok and the Nepal-Swiss Project initially in Dolakha District. Implementation of community forestry has been focused primarily in the Middle Hills region, and has shown a significant increase in coverage in the last decade. In order to replicate the success of community forestry, and modelled on the experience of community partnership in forest management in India (joint forest management), collaborative forest management was initiated in 2002 to involve users in the management of the high-value forests in the Terai region (while retaining a significant degree of control over forest management and ensuring a large portion of the returns from harvesting valuable trees returns to government). Similarly, buffer zone community forestry was established to address the issues of forest-dependent people living in and around national parks and protected areas. Even though the area under leasehold and collaborative forestry has increased substantially, the scale remains fairly insignificant. (See Table 6.1 for the changes in the forest area under different forest management regimes between 2004 and 2015.) Below we present brief summaries for the main tenure regimes in Nepal to illustrate and compare further the tenure rights and instruments and outcomes of the main community-based forest management regimes.

As discussed, **community forestry** is the largest community-based tenure regime in Nepal. Based on the Forest Act 1993, local people form a Community Forest User Group (CFUG) and request the District Forest Office to hand over an identified patch of forest as community forest. Initially community forestry was initiated to restore the degraded Middle Hills and to supply forest products to rural people. As of January 2017, 19,361 CFUGs have been established managing around 1.81 million hectares of forest, over 30 percent of the total forest area of Nepal. The rights that CFUGs enjoy are mainly operational (access and use); and collective choice (management and exclusion). Though there is no defined

Table 6.1 Forest area 2004 to 2015 under various forest tenure regimes in Nepal (in ha)

Tenure regime	2004	2008	2010	2015	2015 %
Community forests	1,153,848	1,229,669	1,381,736	1,898,917	31.1
Leasehold forests	1,677	6,483	8,014	42,835	0.7
Collaborative forests	6,670	3,944	22,929	61,709	1.0
Government-managed forests				3,480,000	57.0
Protected areas and others				416,675	6.8
Private forest				2,056	0.03
Buffer zone CF				198,550	3.3
Total				5,800,000	100

(Source: DoF 2015)

period or legal guarantee of continuity, the management plans are to be renewed within a five- to 10-year period and are renewable.

Despite questions around security of tenure, community forestry has significantly contributed to the protection of forests in the Middle Hills of Nepal (Kanel and Dahal 2008; Pokharel, Tiwari and Thwaites Chapter 3 in this volume). Most of the CFUGs are able to manage forests and generate a group fund. Such funds are being used for local community development activities such as constructing and maintaining school buildings, roads and trails, irrigation facilities and community buildings, etc. (Dahal and Chapagain 2008; Devkota, Thwaites and Race Chapter 5 in this volume). CFUGs have developed policies to support poor and disadvantaged members in a community in a variety of ways. Examples of such policies include provisions for special preferential quota for membership of the Executive Committee and benefit sharing for Dalits and poor, obligatory 50 percent women's participation in all Executive Committees, and providing funds to support income generating activities. While implementation of such provisions remains a challenge, some CFUGs have made efforts to empower and support poor and disadvantaged members; in some cases, CFUGs have funded poor students for their education and health.

Leasehold forest specifically targets the poorest amongst the poor within the community with an aim to improve livelihoods and eventually to alleviate the poverty of those identified poor households living adjacent to degraded forests. The government allocates patches of forest land for 40-year leases to groups of 10 households that have been identified as poor. The group will protect, manage, develop and restore the forest, as well as use products from the forest land. Despite its limited coverage, leasehold forestry has become popular amongst poor households because it has helped to improve forest condition and contributed to the livelihoods of the members. However, due to procedural complexities and limited ability of leasehold groups to access financial and technical assistance from the involved organizations, the progress in terms of scaling up this model has so far been relatively slow compared to the expansion of community forestry. Furthermore, the progress is constrained as most of the forests allocated to leasehold groups are degraded and demand intensive and costly restoration efforts (FAO 2005, 2006).

Collaborative forestry is a relatively new tenure regime in forestry in Nepal. After the revision of the Forest Policy in 2000, the Government of Nepal developed a new regime by adopting a community-based forestry approach specifically to manage large blocks of forest in the Terai region, calling it collaborative forestry. This is a benefit-sharing scheme between the District Forest Office (DFO), local government (District Development Committee) and local forest users. Unlike community forestry, the rights of users to benefits from forest products are limited, as only 50 percent of the benefits from timber harvesting and sale will go to the community group and the decisions are collectively made by the collaborators not the community group alone. The key objectives of collaborative forestry are to meet local demand of users and demand for commercial use of forest products, and to reduce poverty by creating employment while enhancing biodiversity. Local residents living within 5 km from the forest are considered as primary users of collaborative forestry. Collaborative forestry also follows the principles of participatory forest management and has been trying to maintain equity by providing preferential treatment to poor households (discounted price on timber purchase, etc.), and encouraging women to participate in decision-making. As the regime is relatively new, the economic, social and environmental outcomes are yet to be clearly understood.

The scale of **private forests** in Nepal is insignificant so far, making up only 0.01 percent of total forest area, but slowly people are being encouraged towards establishing registered private forests as per the Forest Act 1993. Our observations indicate that some have started doing agroforestry within their private agriculture land or renting other people's land for private forestry with an arrangement of holding tenure over trees. In any case, private forests provide direct economic benefits to the individual, as 100 percent of the benefit goes to the private owner. Global experiences suggest that farmers with abandoned agricultural land are shifting their priority from agriculture farming to tree farming due to a shortage of labour for farming and better income from trees (Gilmour 2015). However, in Nepal, there are a number of disincentives, or barriers to farmers planting trees on private land, including legal restrictions on growing and selling certain species of trees on private land, and also VAT (value added tax) is imposed on commercial sale of the forest products. Further, registration of private forests is a legal requirement to have rights to harvest timber and to sell timber commercially. Yet the process of registration of private forest at the District Forest Office is complex, presenting a further barrier to private landholders growing trees on their farmland. Unlike some other community-management models, no comprehensive studies have been undertaken to find out the impacts of private forests. Nevertheless, increasing interest from people indicates that private forests are perceived to contribute to livelihoods and income through employment generation and sale of forest products (Sikor et al. 2013). In addition, private forests could help in maintaining local environmental condition.

State-managed forest is the most widespread tenure regime, covering more than 60 percent of the total forest area of Nepal, including forests under government management, protected areas, and parks and wildlife reserves. Research by Paudel, Banjade and Dahal (2008) has revealed that the condition of forests under state management is poor and gradually declining due to illegal activities and poor silvicultural practices. The role of local communities, indigenous people and private sector actors is completely ignored within the state-managed forests. Government alone is responsible to protect, manage and use the forests. In many cases, the government has not developed management plans to maintain forests. None of the rights under the tenure bundle is given to the local people, therefore creating a potential conflict between forest authorities and communities.

In Nepal, the reform of forest tenure from a state-controlled regime to community-based tenure has significantly improved forest condition mainly across the Middle Hills (Larson, Barry and Dahal 2010; Pokharel et al. 2008). Numerous studies have been undertaken to assess the change in forest condition after the change to community-based forestry tenure (e.g. Branney and Yadav 1998; Gautam et al. 2003). Based on review of published work, Pokharel, Tiwari and Thwaites (Chapter 3 this volume), have shown that community forestry tenure has resulted in an increase in forest cover and density and improvements in a variety of forest condition measures, particularly across the Middle Hills of Nepal. In many cases, biodiversity has also been shown to increase under community forestry. Similarly, the study on leasehold forests by Singh and Chapagain (2006) indicated that growth of grass, regeneration of saplings, plant species diversity and development of overall greenery have increased with decline in *khoria* cultivation (slash and burn). An earlier study by Singh and Shrestha (2002, cited in Singh and Chapagain 2006) found vegetation cover had increased from an average of 32 percent to 78 percent in leasehold forests within a six- to seven-year period. Government-managed forests have shown poor performance in terms of maintaining forest condition and make no contribution to the livelihoods of forest-dependent people (Pokharel et al. 2008). However, though small in scale, private forests have contributed to the livelihoods of local people and helped improve the local environment (Paudel, Banjade and Dahal 2008). Table 6.2 presents a summary of the bundle of rights granted under various tenure categories and their outcomes, and identifies the legitimate right holders.

Table 6.2 highlights how the outcomes arising from forest tenure under community management are better than those from state management in terms of forest condition and benefits to users. A number of studies have indicated a link between the protection and active management through adoption of silvicultural practices in community-managed forests and improvements in the condition of those forests, which can be compared to the poor condition of state-managed forests where silvicultural operations are not practiced. It could be argued that the willingness to invest in labour-intensive management following silvicultural principles is influenced by the bundle of rights that are held under community management, as benefits of the investment are more likely to be captured by the forest users. However, as highlighted by Pokharel, Tiwari and Thwaites

Table 6.2 What rights and who holds those rights under different management regimes?

		Who has rights?								
Management Regimes →	State management			Collective management					Private management	
What rights ↓	Govt. managed	Protected forests	National parks and reserves	Community forests	Leasehold forests	Collaborative forests	Buffer zone forests	Religious forest	Private forests	
Access	Govt.	Govt.	Govt.	Community	Community	Community and government	Community	Community	Individual	
Use	Govt.	Govt.	Govt.	Community	Community	Community and government	Community	Community	Individual	
Management	Govt.	Govt.	Govt.	Community	Community	Community and government	Community	Community	Individual	
Exclusion	Govt.	Govt.	Govt.	Community	Community	Community and government	Community	Community	Individual	
Alienation	Govt.	Govt.	Govt.	Government	Government	Government	Government	Government	Individual	
Duration	Unlimited	Unlimited	Unlimited	Tenure period not defined by laws but regulated by the management plan valid for five or 10 years	40 years, renewable for another 40 years	Unlimited, but regulated by the management plan valid for five years	Unlimited but regulated by the management plan valid for five years	Unlimited but regulated by the management plan valid for five years	Unlimited	
Extinguishability	Not applicable	Not applicable	Not applicable	Limited compensation Govt. can revoke	Limited compensation, Govt. can revoke	Not applicable	Not applicable	Not applicable	Actual compensation, Govt. cannot claim private property	

(Continued)

Table 6.2 Continued

		Key features and outcomes						
Management Regimes →	State management			Collective management					Private management
What rights ↓	Govt. managed forests	Protected forests	National parks and reserves	Community forests	Leasehold forests	Collaborative forests	Buffer zone forests	Religious forest	Private forests
Benefits and revenue management	No recognition of people's rights over forest. 10 percent revenue to local government. Cost to be allocated by national treasury.			Users have rights to fix royalties and sell products Restriction on commercial sale 100% benefit goes to community. 15% charge on commercial sale 25% of income to forest development	Existing trees are not allowed to be used for commercial sale as forest product harvesting is restricted 100% benefits to community	Use of lops and tops and 50% from sale of products goes to local community	30–50% of park benefits to be used on buffer zone management, forest users have no right to sell the products	Recognition of traditional use of forests, restriction on sell of forest products for commercial use	VAT is imposed, need to pay land revenue, growth and sale of certain species is restricted 100% benefits to individual

Outcomes as of 2016	Poor forest condition, illegal logging, no silviculture practices, no rights to access and use forests by local people	Improved forest condition, limited economic benefits to users, increased exercise for equity and democracy	Improved forest condition, economic benefits to users	Multi-stakeholder participation, improved forest condition	Limited incentives to local people, more restriction resulted in conflict between govt. and local users	Improved forest condition, no rights to use forest products.	Greater economic impact as all benefit goes to individual household Support maintaining local environment

(Source: author's compilation, 2016)

(Chapter 3 this volume), there is tension between silvicultural operations to deliver improved supply of forest products and a more 'passive' approach to management in a protection-oriented management regime. As noted by Larson and Dahal (2012) to date, community-based forestry has had limited success in harnessing economic benefits, particularly in terms of the distribution of those benefits amongst the local communities who protect and manage the forests. Devkota, Thwaites and Race (Chapter 4 this volume) report that silvicultural practices are often poorly applied, and there is a need to improve silvicultural practices to improve economic and livelihood benefit to local users. Restrictions on growing certain species are also imposed on private forests, thus reducing the motivation to establish private forests in Nepal (Larson and Dahal 2012; Paudel, Banjade and Dahal 2008).

Informing the global and regional practices and learning for Nepal

Just as in Nepal, forest tenure at the global level has followed a similar pattern, with evolution of reform from state-controlled to more participatory community-controlled tenure models. In many countries, land and forest were initially managed through customary practices by local communities and indigenous people, before coming under increasing state control. This section considers the trends in forest tenure reform unfolding at the global and regional levels in order to understand how Nepal's policies are informed by the global and how the global trend is informed by Nepal's experience.

In the countries where European colonies existed, national resources were appropriated for the benefit of the rulers and it became a norm for the state to be the owner and manager of natural resources including land and forests, a situation which continues to be the dominant regime in forestry across the globe. While Nepal was never a colony, its policies in general are influenced by the British-Indian policies and colonial legacies have greatly impacted Nepal's forestry sector. Under state control, in Nepal and elsewhere, forest-dependent people were seen as a problem, incapable of exercising rights in managing forests, and thus were policed. States exploited forests applying different models of revenue generation by contracting concessions to private contractors. Furthermore, in situations where state control excluded indigenous systems of management, in the expectation of conserving the forests, strategies for controlling the access of forest-dependent communities to forests and forest resources were adopted, inevitably resulting in a 'tragedy of the commons' situation (Hardin 1968), in which communities are forced to extract resources illegally and unsustainably. In general, state-controlled forest systems have performed poorly in situations where local people depend on forest resources, resulting in an alarming rate of forest degradation and increase in poverty amongst forest-dependent people (Yasmi et al. 2010).

The rethink of international development discourse in the 1970s and 1980s sought to decentralize powers to local communities, which also impacted Nepal's forest policies, culminating in the pioneering community forestry programs discussed above (Esman and Uphoff 1984; Adhikari 2015). According to RRI figures (RRI 2014; FAO 2010), the total global forest area under the legal ownership or control of indigenous people and local communities has gradually increased from 383 million hectares (11 percent) in 2002 to 511 million hectares (15.5 percent) in 2013. With about double (31 percent in 2015) the share of forests managed by the community organizations and groups, Nepal seems to be doing well in global terms. Despite a shift in policies in the late twentieth century involving tenure reforms and devolution of rights to local communities and private entities, states continue to hold ownership as well as management rights of most of the world's forests (73 percent, compared to 60 percent in Nepal) (Figure 6.1).

Analysis of tenure trends across geographic regions shows considerable variation in the status of forest tenure and the level of distribution across various tenure categories (Table 6.3). There are substantial differences in cultural and economic histories, legal traditions and pressures on forest resources across the different regions. Many countries in Latin America have implemented significant reforms in forest tenure, recognizing the rights of local communities and indigenous peoples (IPs), and forest tenure is more balanced and widely distributed across all tenure categories (public and private), with a significant change evident between 2002 and 2013. In Africa, forest tenure is predominantly (around 93 percent) under state control and management with very limited recognition of statutory community rights. Recognition of the rights of communities and IPs to forest lands and resources has taken longer in most countries in Africa than in Latin America and Asia.

The situation in terms of share of community-based forest tenure regime is slightly better in Nepal, where almost one-third of forest land is either owned

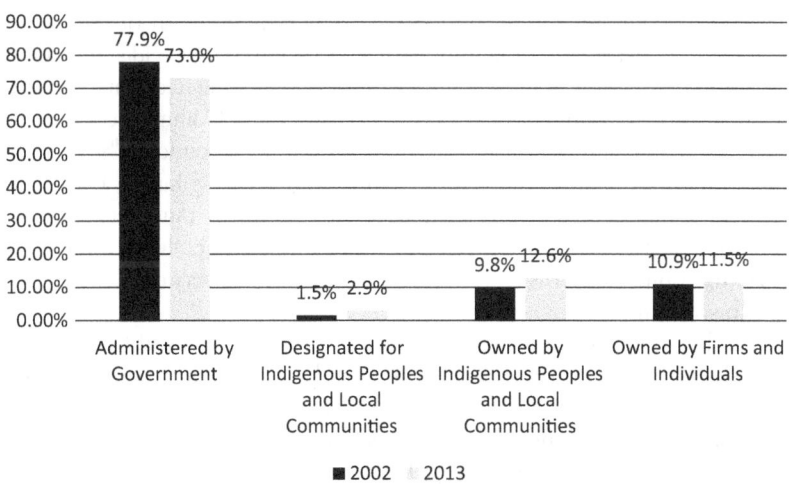

Figure 6.1 Global trend of forest tenure, 2002 and 2013, by percentage
(Source: RRI 2014)

Table 6.3 Regional trends of statutory forest tenure (as percent of forest area)

Tenure categories	Nepal		Asia		Africa		Latin America	
	2002	2013	2002	2013	2002	2013	2002	2013
Public								
Administered by government	81.6	68.0	67.4	60.9	95.5	93.7	61.1	42.9
Designed for IPs and communities	18.4	32.0	3.5	6.0	4.2	5.9	2.0	6.2
Private								
Owned by IPs and communities	0.0	0.0	26.9	30.6	0.3	0.3	22.4	32.9
Owned by firms and individuals	0.0	0.0	2.2	2.5	0.0	0.0	14.5	17.9
			100%	100%	100%	100%	100%	100%

(Source: RRI 2014)

by or is designated for communities and IPs, compared to the overall situation in Asia, and much better than in Africa where the figure is only 6 percent. There remains a significant difference in the type of tenure and the level of tenure security across countries. Central governments in most countries in Asia have been creating space for non-state actors, mainly local communities and indigenous people, in the protection and management of forest resources. In many cases, governments have transferred authority to local communities, providing certain rights to access, use and manage forests. As of 2013, almost 31 percent of forests in Asia are owned by IPs and local communities and a further 6 percent are under community management and control.

Forest tenure reform in China and Vietnam has shown that tenure security (private ownership) is helping to achieve better economic and environmental outcomes (RRI 2014; Yasmi et al. 2010). The long duration of tenure period, such as for 70 years in China and 50 years in Vietnam, provides greater confidence and motivation to individual households in making necessary investments and realizing the benefits from forest management. Community-based forest tenure such as community forestry in Nepal has been found to be effective in improving forest condition collectively (Chapter 3 this volume), but as already mentioned, has faced difficulty in sharing benefits to each member of the forest user group as establishing a system for redistribution of collective income, and providing equitable access to resources has proven to be very challenging (Chapter 4 this volume).

The global and regional trend of transfer of tenure from state to other tenure holders (communities, indigenous people and private households) indicates that there is an increasing understanding amongst governments that forests management is not the sole responsibility of the state, but rather that it should be managed collectively. However, contrary to this understanding, governments in many countries remain reluctant to share revenue from the forest; hence strong tenure security could be considered to be a pre-requisite for legitimate non-state tenure holders, mainly those who protect and manage the forests, to claim benefit.

Global and regional learning inform us that Nepal's community forestry could follow the example of China and Vietnam in seeking to overcome some of the obstacles to achieving desired outcomes, if there is secured tenure over land and forests in favour of communities and indigenous people, and if the state allows local people to generate revenue through harvesting and sale of timber and non-timber forest products (based on technical recommendation as agreed in the management plan) from community forestry without any restriction.

Conclusion: future considerations

Over the last four decades, significant and radical changes have been taking place in the statutory forest tenure arrangement in Nepal. Community-based forest tenure has pioneered the devolution of rights and responsibilities from the state to local communities. Even though community forestry in Nepal does not provide alienation and extinguishability rights to the local community, it is often regarded as the key example of decentralized and participatory forest management due to its large coverage, long history, level of autonomy and significance to local livelihoods. In its history of over 40 years of implementation, community forestry has evolved from a small, localized and experimental forest management model to a major national program involving millions of rural people across the country.

Nepal's community forestry has drawn from the global discourse of participatory development and local history of customary systems of collective management of natural resources, decentralization and local governance. Nepal's case has clearly demonstrated that, given appropriate tenure rights, people are capable of managing their forests and improving the forest condition and ecological diversity. Nepal's initiative has inspired many other countries to learn from Nepal and adopt their own community-based tenure policies and Nepal has also been able to learn from experiences in other countries. Studies so far suggest that increasing forest coverage and keeping forests intact does not necessarily result in improved livelihoods for all members of community forestry groups (see Chapter 4). In many instances, community forestry can have a negative impact, as the traditional rights of forest dwellers, chauri (yak/domestic cow cross) herders, honey collectors and agricultural tool makers (see Chapter 8 for the case of blacksmiths) are ignored and their access to livelihood resources is restricted, forcing them to change their traditional livelihood options. Hence the bundle of rights to manage at the collective level should include a provision to protect the rights of forest-dependent people to access forest resources.

There are still a number of unresolved questions pertaining to community forestry in Nepal. How secure are the current tenure arrangements? Who owns the carbon rights for community forests and how will the benefits from trading of carbon be distributed? These are some of the vital issues that require further research and evidence before community forestry in Nepal can be claimed as successful. Finally, in order to achieve social, economic, and environmental objectives of forest management, community forestry tenure in Nepal needs to be progressed as a part of a holistic and integrated society-wide reform agenda to support empowerment of marginalized groups, particularly women and the poor, and forests should be used for economic benefits and prosperity rather than only protecting forests to meet environmental objectives.

References

Adhikari, B. and Lovett, J.C. 2006, Transaction costs and community-based natural resource, *Journal of Environmental Management*, vol. 78, pp. 5–15.

Adhikari, K.P. 2015, Rural development policies, CBOs and their sustainability in Nepal, in A. Adhikari, G.P. Dahal, K. Subedi, I. Mahat and B. Regmi (eds), *Sustainable Livelihood Systems in Nepal: Policies, Practices and Prospects*, IUCN and CFFN, Kathmandu, Nepal, pp. 237–263.

Agrawal, A. and Ostrom, E. 2001, Collective action, property rights, and decentralization in resource use in India and Nepal, *Politics and Society*, vol. 29, no. 4, pp. 485–514.

Branney, P. and Yadav, K.P. 1998, *Changes in community forest condition and management 1994–1998: Analysis of Information from the forest resource assessment study and socioeconomic study in the Koshi Hills, Nepal*, Project Report No. G/NUCFP/32, Nepal UK Community Forestry Project Kathmandu, Nepal.

Clark, I. 2000, *Governance, the State and Industrial Relations*, Routledge Publications, London, UK.

Dahal, G.R., and Chapagain, A. 2008, Community forestry of Nepal: Decentralized forest governance, in C. Colfer, G. R. Dahal and D. Capistrano (eds), *Lessons From Forest Decentralization: Money, Justice and Quest for Good Governance in Asia and the Pacific*, Earthscan Publications, London, UK.

DoF 2015, *Hamro Ban-2072*, Department of Forests, Babar Mahal, Kathmandu, Nepal.

Esman, M. and Uphoff, N. 1984, *Local Organizations: Intermediaries in Rural Development*, Cornell University Press, London, UK.

FAO 2005, *Global Forest Resource Assessment*, Food and Agriculture Organization of the United Nations, Rome, Italy.

FAO 2006, *Understanding forest tenure in South and Southeast Asia*, Forestry Policy and Institutions Working Paper No. 14, Food and Agriculture Organization of the United Nations, Rome, Italy.

FAO 2010, *Global Forest Resources Assessment, Country Report, Nepal*, Food and Agriculture Organization of the United Nations, Rome, Italy.

FAO 2011, *Reforming Forest Tenure- Issues, Principles and Process*, FAO Forestry Paper Series 165, Food and Agriculture Organization of the United Nations, Rome, Italy.

FAO 2015, *Global Forest Resource Assessment*, Food and Agriculture Organization of the United Nations, Rome, Italy.

Fisher, R.J. 1989, *Indigenous system of common property forest management in Nepal*, Working Paper No. 18, East West Environment and Policy Institute, East West Center, Honolulu, HI.

Gautam, A.P., Webb, E.L., Shivakoti, G.P. and Zoebisch, M.A. 2003, Land use dynamics and landscape change pattern in a mountain watershed in Nepal, *Agriculture, Ecosystems & Environment*, vol. 99, no. 1/3, pp. 83–96.

Gilmour, D.A. 2015, *Unlocking the wealth of forests for community development: Commercializing products from community forests*, Presentation at IUFRO 3.08 Small Scale Forestry Conference, Sunshine Coast, Australia, 11–15 October.

Gilmour, D.A., Durst, P.B. and Shono, K. 2007, *Reaching Consensus, Multi-Stakeholder Process in Forestry: Experiences From the Asia-Pacific Region*, Food and Agriculture Organisation, Bangkok, Thailand.

Hardin, G. 1968, The tragedy of the commons. *Science*, vol. 162, pp. 1243–1248.

Höfer, A. 1979, *The Caste Hierarchy and the State in Nepal: A Study of the Muluki Ain of 1854*, Universitätsverlag Wagner, Innsbruck, Austria.

Kanel, K. and Dahal, G.R. 2008, Community forestry policy and its economic implications: An experience from community forestry of Nepal, *International Journal of Social Forestry*, vol. 1, no. 1, pp. 50–60.

Larson, A., Barry, D. and Dahal, G.R. 2010, *Forests for People: Community Rights and Forest Tenure Reform*, Earthscan, London, UK.

Larson, A.M. and Dahal, G.R. 2012, Forest tenure reform: New resource rights for forest based communities? *Conservation and Society*, vol. 10, no. 2, pp. 77–90.

Malla, Y.B. 2000, Impact of community forestry on rural livelihoods and food security, *Unasylva*, vol. 51, no. 202, pp. 37–45.

Paudel, N.S., Banjade, M.R. and Dahal, G.R. 2008, *Devolution challenges in Nepal's community forestry in the context of emerging market opportunities*, Policy Brief, Forest Action, Nepal.

Pokharel, B., Branney, P., Nurse, M., and Malla, Y.B. 2008, Community forestry: Conserving forests, sustaining livelihoods, strengthening democracy, in H. Ojha, N. Timsina, C. Kumar, B. Belcher and M. Banjade (eds), *Communities, Forests and Governance: Policy and Institutional Innovations From Nepal*, Adroit, New Delhi, India, pp. 55–91.

Regmi, M.C. 1963, *Land Tenure and Taxation in Nepal*, Vol. 1, Institute of International Studies, University of California, Berkley, CA.

Regmi, M.C. 1977, *Land Ownership in Nepal*, Adroit Publishers, New Delhi, India.

Ribot, J.C. 2002, *Democratic Decentralisation of Natural Resources: Institutionalising Popular Participation*, World Resources Institute, Washington, DC.

RRI 2012, *What Rights? A Comparative Analysis of Developing Countries' National Legislation on Community and Indigenous Peoples' Forest Tenure Rights*, Rights and Resources Initiative, Washington, DC.

RRI 2014, *What Future for Reform? Progress and Slowdown in Forest Tenure Reform Since 2002*, Rights and Resources Initiative, Washington, DC.

Sikor, T., Gritten, D., Atkinson, J., Huy, B., Dahal, G.R., Duangsathaporn, K., Hurahura, F., Phanvilay, K., Maryudi, A., Pulhin, J., Ramirez, A., Win, S., Toh, S., Vaz, J., Sokchea, T., Marona, S. and Yagio, Z. 2013, *Community Forestry in Asia and the Pacific: Pathway to Inclusive Development*, Centre for People and Forest-RECOFTC, Bangkok, Thailand.

Singh, B.K. and Chapagain, D.P. 2006, Trends in forest ownership, forest resource tenure and institutional arrangements: Are they contributing to better forest management and poverty reduction? Community and leasehold forestry for the poor: Nepal case study, in FAO (ed.), *Understanding Forest Tenure in South and Southeast Asia*, Forest policy and institutions working paper 14, pp. 115–152, Food and Agriculture Organization of the United Nations, Rome, Italy.

Singh, B.K. and Shrestha, B. 2000, *Group site information report of leasehold groups*, Hills Leasehold Forestry and Forage Development Project, Kathmandu, Nepal.

Whelpton, J., 2005, *A History of Nepal*, Cambridge University Press, Cambridge, UK.

White, A. and Martin, A. 2002, *Who Owns the World's Forests? Forest Tenure and Public Forest in Transition*, Forest Trends, Washington, DC.

Yasmi, Y., Broadhead, J., Enters, T. and Genge, C. 2010, *Forestry policies, legislation and institutions in Asia and Pacific: Trends and emerging needs for 2020*, Working Paper No. APFSOS II/WP/2010/34, Food and Agriculture Organization, Bangkok, Thailand.

Yasmi, Y., Kelley, L. and Enters, T. 2010, *Conflict over forests and land in Asia: Impacts, causes and management*, RECOFTC Issues Paper, RECOFTC, Bangkok, Thailand.

7 Community forestry and pro-poor climate change adaptation

A case study from Nepal

Popular Gentle and Richard Thwaites

Introduction

Climate change is a global issue that is affecting the lives of people in Nepal today. The implications of climate change tend to be more severe in developing countries where the majority of the population are dependent on rain-fed agriculture options for their livelihoods and have limited capacity to adapt to changing weather patterns and seasonal variations (MoE 2010). Climate change has created additional burdens for developing nations, particularly for those already suffering from poverty and increasingly unsustainable livelihoods. While the impacts of climate change have been severe over the past two decades, projected scenarios indicate a worsening future.

The fragile livelihood and environmental base in rural Nepal result in significant adverse effects from climate change. Ebi et al. (2007) show that significant warming has been experienced at high altitudes in the hills and mountains of Nepal. In mountainous areas, temperature increases have been higher than global averages over the last 100 years. Observed impacts of climate change in Nepal include glacial retreat, drought, erratic rainfall and the unpredictable onset of monsoon seasons, storms and landslides (MoE 2010). These occurrences have resulted in crop failure, decreased food and livelihood security, induced water scarcity, increased prevalence of some human diseases and increased income insecurity. More than 60 percent of the cultivated area of Nepal relies entirely on monsoonal rainfall, and increasingly unpredictable weather patterns are affecting production of staple crops. In rural Nepal, the poorest people are most vulnerable to changing climate, as they depend on dryland agriculture, have limited livelihood options and low adaptive capacity. Households dependent on the natural resource base have been found to be more vulnerable than those whose livelihoods were based in other sectors (MoE 2010).

Resource dependent communities are historically experienced in managing weather-dependent natural resources and have developed various adaptation practices to reduce their risks and vulnerabilities. However, coping and adaptation strategies applied by local communities may not be adequate due to lack of information, knowledge and resources in the face of increasing climate change-induced vulnerabilities. Climate change has implications for natural resource-based livelihoods and the projected risks are more profound where live-lihoods depend primarily on weather-sensitive natural resources. Local institutions can play an important role in responding to climate change impacts and in supporting local communities to enhance their capacity to adapt at individual, household and community levels. According to Adger (2003, p. 387), "adaptation is a dynamic social process [and] the ability of societies to adapt is determined, in part, by the ability to act collectively". Adger (1999) differentiates individual and collective vulnerabilities with their causes and indicators, listing the causes of individual vulnerability as relative and absolute poverty, entitlement failure and resource dependency; and of collective vulnerability as levels of infrastructure and development, institutional and political factors, insurance, and formal and informal social security.

In supporting local community adaptation to climate change, local institutions can influence the distribution of climate risk by organizing incentives for households and community-level adaptations, and mediating external interventions suited to the local context (Agrawal 2010). The most widespread and well-established local institutions in rural Nepal are the Community Forest User Groups (CFUGs). These have a progressive mandate (e.g. require fund allocation for poor people, have a focus beyond local forests) and have potential to contribute to climate change adaptation by providing ecological goods and services, socio-economic benefits and a 'safety net' for poor people (Pokharel and Byrne 2009). However, concerns have been raised that CFUGs do not always deliver on their progressive mandate (e.g. problems associated with equity issues) and may not provide a viable safety net for poor people confronted by the challenges of climate change (Acharya and Gentle 2006).

This chapter explains how the livelihoods of rural communities in Nepal are influenced by climate change, describes vulnerability under threat from climate change and considers how people with different well-being status are addressing the changes and the status of existing support mechanisms for these communities. In addition, the potential roles of local institutions, especially CFUGs, in developing and implementing pro-poor climate change adaptation strategies in the rural areas of Nepal are discussed.

Local institutions, collective actions and climate change adaptation

There are many cases in which rural communities have enhanced adaptive capacity and organized collectively to manage climate risks using their local institutions in the form of social networks, capital, norms and traditions. The formation and functioning of social networks is linked with the response capacity of the system because collective actions can mediate collective risks and enhance adaptive capacity to climate change.

Many studies emphasize the potential for rural institutions to facilitate adaptation to climate change and strengthen adaptive capacity in the local context (Adger 2000; Agrawal 2010). Based on research undertaken in Mexico, Eakin

(2005, p. 1936) argues that the sensitivity of farmers to climatic impacts and their capacity to manage climatic risk depends on "how they organize their livelihoods in confronting institutional change". Similarly, Robledo, Fischler and Patiño (2004) found that community organizations were successful in developing adaptation strategies and building resilience of hill communities in Bolivia through ecosystem management and restoration activities including rehabilitation of watersheds, agro-ecology and the forest landscape. Moser (1996) and Narayan-Parker (1997) have also shown that communities with effective civic associations and social networks are likely to be more successful in coping with adverse situations caused by climate change.

According to Agrawal and Perrin (2008) climate change adaptation is a local process, and its effectiveness depends on local and external institutions and the incentives they craft for individual and collective action. This is because:

> Institutional arrangements structure risks and sensitivity to climate hazards, facilitate or impede individual and collective responses, and shape the outcomes of such responses.
>
> (Agrawal 2010, p. 174)

Agrawal and Perrin (2008) outline three ways by which institutions influence the livelihoods and adaptations adopted by rural communities.

1 Institutions structure the distribution of climate risk impacts. Although the physical and structural characteristics of the hazard determine how particular social groups will be affected by climate hazards, local institutions with good governance can play an important role in reducing the ill effects and distribution of risks.
2 Institutions constitute and organize the incentive structures for household and community-level adaptations. As institutional incentives basically define the cost of collective action and the extent of transaction costs, they also help to determine whether adaptation responses will be managed by individuals or managed collectively.
3 Institutions mediate external interventions into local contexts which ultimately stimulate the adaptation by articulating social and political processes. Local institutions can play a crucial role in the design and implementation of external adaptation-related interventions such as providing knowledge, information, awareness, technical assistance, skills and financial support.

Potential and contribution of CFUGs in climate change adaptation

Prior research (e.g. Agrawal and Perrin 2008) has recognized the current role and potential of local institutions in climate change adaptation. CFUGs in Nepal have been shown to function in coordination with various formal and informal local institutions from the state, civil society and private sectors. The existing knowledge, capacity and experience of CFUGs is very low in designing and implementing climate change-related activities and in addressing the needs of local communities for climate change adaptation initiatives. Although the plans and activities of many CFUGs have no explicit provisions for climate change adaptation, they have been found to be able to make a useful contribution to climate change adaptation by enabling communities. The National Adaptation Program of Action (NAPA) framework prepared by the Government of Nepal (MoE 2010) has recognized local institutions such as CFUGs, farmer groups and irrigation groups as important agencies for implementation of local-level adaptation activities, and has declared its intention to allocate 80 percent of total adaptation funds to these institutions.

According to various literature (Table 7.1), CFUGs in Nepal provide well-established networks with locally governed rules, norms and values to mobilize local communities and manage common pool resources. The major roles and functions of CFUGs to support local communities in climate change adaptation can be highlighted as: (i) income and employment generation; (ii) protecting and managing forest commons; (iii) supply of forest products and contribution to rural livelihoods; (iv) delivering environmental services such as protection and conservation of soil and water resources; (v) contributing to infrastructure development and community development activities; (vi) targeting pro-poor investment and contributing to poverty reduction; (vii) managing livestock grazing; (viii) awareness raising, capacity building and leveraging social capital; and (ix) supporting communities in disaster risk reduction and emergency relief. Table 7.1 summarizes some strengths and limitations of CF for climate change adaptation in Nepal.

Table 7.1 Strengths and limitations of community forestry for climate change adaptation

Areas	Strengths	Limitations
Scale and coverage	One of the largest networks of civil society organizations in Nepal Nationwide coverage (in all 75 districts) and rooted in rural areas where adaptation need is high Incorporating about 35% of population of the country and over 60% of households in rural areas.	Limited coverage of CF in high Mountain and Terai regions Access to CF resources and equitable distribution of benefits to poor has been criticized and challenged by several studies
Livelihood assets	**Natural assets:** Managing 1.81 million ha forest area with forests, rivers, shrubs, non-timber forest products, soil and water conservation, rich biodiversity, grazing land and forest products as agriculture inputs. Land allocation for pro-poor income generating and livelihoods improvement activities	Low participation of women and Dalits in decision-making in CFUGs Nominal amount of CFUG fund allocation for pro-poor activities Corruption or misuse of resources in some cases

Areas	Strengths	Limitations
	Financial assets: Collecting and mobilizing over USD10 million annually, CFUG fund mobilization as a 'safety net' for the poor. CF in Nepal has over 180 million tonnes of sequestered carbon **Human assets:** Successfully mobilized over 2.18 million households and about 195,000 committee members for day-to-day decision-making, contributes large amount of CFUG funds as salary of school teachers, scholarships for poor children, skill-based training, employment in forest protection and management as well as in community development activities **Physical assets:** CFUG offices and facilities, CFUG funds provide a major source of infrastructure construction (such as drinking water, road, electrification, bridge, school building) in remote areas **Social assets:** Established practices of general assembly, election of committees and mobilization of large population of users. Well-established and vibrant network of FECOFUN at VDC, district and national levels	Greater orientation and motivation towards commercialization of resources than subsistence needs Limited awareness and knowledge of climate change and adaptation
Mandate and policies	Objectives of CF: protection, management and utilization of forest resources, poverty reduction, livelihoods improvement Over 30 years of established credibility and legitimacy from community, government, donor and international community; high investment program Pro-poor and equitable policies (mandatory provision of well-being ranking of all households and pro-poor targeting; provision to allocate 35% of income for pro-poor activities, allocation of CF land for pro-poor livelihoods, proportionate representation of gender, caste, ethnicity in decision-making) Autonomous organization with democratic functioning NAPA document recognized CFUG as one of the major implementing agencies at the local level.	Climate change adaptation may be considered as a drift from original mandate Barriers to practicing pro-poor and affirmative actions Ensuring equitable benefit sharing in the context of elite dominance in benefit sharing No functional relationship with Ministry of Environment, the leading ministry in climate change adaptation

(Source: DoF 2011; DoF 2017; GoN 2009; Kanel 2004; MoE 2010; Pokharel and Byrne 2009)

Case study in four CFUGs

The findings of this chapter are based on case study research conducted in the Lamjung District of Nepal from 2011 to 2013 by Gentle (2014). The district is located in the Middle Hills of western Nepal, where most people depend on subsistence agriculture, farming has a strong relationship with forestry, and CFUGs are well established as institutions involving forest management, community development and social networks. This district has been rated as one of the districts in Nepal most vulnerable to climate change (MoE 2010). Four CFUGs were selected for this research, coming from four different locations. The locations, based on Village Development Committees (VDCs), were purposefully selected for this research in consultation with district-level government and NGO stakeholders considering: (i) representation of physiographic, ecological, socio-economic and cultural diversity of the district; (ii) VDCs outside the Annapurna Conservation Area; (iii) VDCs with a reasonable number of CFUGs, as CFUGs are proposed as research units of this research; and (iv) accessibility of the VDCs to allow the research activities to be carried out within the given time (Table 7.2).

This research adopted a multi-scale and mixed methods approach involving predominantly ethnographic qualitative methods of data collection complemented by some quantitative methods to highlight individual actions and social implications. Data collection carried out at community, district and national levels in Nepal in 2012 and 2013 included: in-depth interviews (n=74 interviewees); focus-group discussions (FGD) (n=11 events with 117 participants); and participant observation, supported by a household-level survey (n=133 respondents, 33 percent of total households in the research area). In addition, four events using the Climate Vulnerability and Capacity Analysis framework (Daze, Ambrose and Ehrhart 2009) assisted in understanding the differential impacts

Table 7.2 Characteristics of CFUGs selected for research

Name of the CFUG	Major caste and ethnicity	Location (VDC)	Altitude (metres)	Number of households	Year CF handed over to the community
Kataharbari	Bahun, Chhetri, Dalits	Tarkughat (downstream)	< 500	80	1995
Raniswanra Sankharpakha	Bahun, Chhetri, Dalits	Archalbot (downstream)	< 500	130	1996
Chisapani	Bahun, Chhetri, Gurung, Dalits	Bahundanda (upstream)	> 1,000	60	2005
Manasalu	Gurung, Dalits	Ghermu (upstream)	> 1,000	120	2003

(Source: DFO and CFUG records)

Table 7.3 Results of well-being ranking in case study CFUGs

CFUGs	Households according to well-being status				
	Well-off	*Medium*	*Poor*	*Very Poor*	*Total households*
Kataharbari	29 (36%)	16 (20%)	18 (23%)	17 (21%)	80
Raniswanra	54 (42%)	41 (32%)	28 (22%)	7 (5%)	130
Chisapani	18 (30%)	28 (47%)	9 (15%)	5 (8%)	60
Manasalu	21 (16%)	70 (53%)	27 (20%)	14 (11%)	132
Total	122 (30%)	155 (39%)	82 (20%)	43 (11%)	402

of climate change on people's livelihoods, effects of hazards on poverty, and the roles of institutions and policies used for climate change adaptation.

Participatory well-being ranking was conducted in each research site to categorize the research households into four well-being strata (well-off, medium, poor and very poor) by applying locally identified criteria of well-being (Mosse 1994). The well-being groups, in turn, were considered as a basis of analysis. The resultant ranking has been found to be useful as a means of socio-economic stratification of households as applied by different researchers in Nepal and abroad (Gentle and Maraseni 2012; Richards, Maharjan and Kanel 2003). Participants in the ranking exercise identified key criteria relevant to their own communities such as: landholding size, quality of land, quality of house, food sufficiency, income sources, status of money lending and borrowing, number and quality of livestock, as well as educational and social status of the family (Table 7.3).

The FGDs were conducted with women and marginalized communities and found to be very effective to understand opinions of marginalized groups who usually hesitate to express their opinions in a mixed group in a hierarchical society.

Qualitative data was analyzed using NVivo 10 through a thematic hierarchical approach, and the quantitative data was analyzed using SPSS using statistical tests such as Pearson's chi-square test and Kruskal-Wallis chi-square and rank sum test.

Case study findings

Differential impacts of climate change

Climate change has differential impacts on the livelihoods of communities, varying largely according to the well-being status of the households. The findings are based on how people perceive the impacts of climate change, how hazards are differentially impacting livelihoods, and how livelihoods of those of different well-being status are exposed to and sensitive to climate hazards and their capacity to adapt to the impacts. Although perceived climate change impacts varied according to location, gender of head of the household and occupation, well-being status was found to be the most significant factor. The details about differential impacts of climate change, as part of this research, have been published in a separate article (Gentle et al. 2014).

Differential impacts based on well-being status were found to be associated with the socio-economic and political context of the research area and society. For example, analysis of the impacts of erratic rainfall showed that the impacts were greater in the rain-fed and upland farms than in the irrigated lowlands. A further analysis showed that households relying on rain-fed uplands were mostly poor, and there was a significant association between well-being status of households and access to irrigation facilities for growing primary crops, with very poor and poor lacking or having very limited irrigation facilities.

A further example of the influence of socio-economic and political context was the experience of landslide vulnerability by households of different well-being groups. Although the increasing occurrence and impacts of landslides was experienced by almost all research participants, the impacts were differentially experienced by the households according to their well-being status. Results showed that the impact of landslides on their farmlands and communal lands was reported by well-off and medium households. However, landslide vulnerability for very poor and some poor households was also on their residential areas in addition to the impacts on farm and communal lands. More than two-thirds of very poor and almost all Dalit families were historically living in landslide-prone areas in all research villages (e.g. Figure 7.1). The households were mostly landless, indebted and had a poor social network. The landslide, as a hazard, was not only impacting the physical resources of the households living in the landslide-prone area, it was also causing threats and psycho-social stress to these families. Many very poor households were compelled to stay in public places during the rainy season

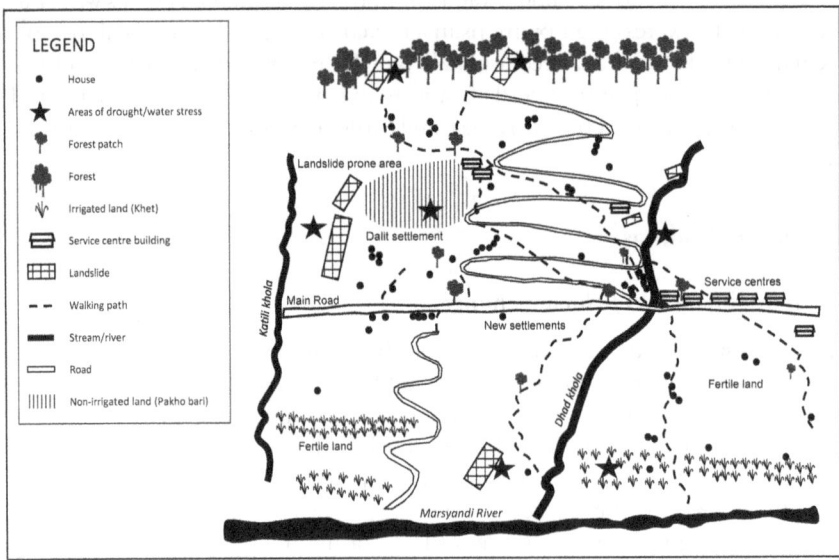

Figure 7.1 Participatory hazard and livelihoods resource map of Kataharbari CFUG[1]

because of the threat of landslides. The analysis showed that the impact of landslides on livelihoods was related to historical marginalization processes, where powerless households and communities were compelled to live in marginal lands.

A third example was the differential experiences of water scarcity, water stress and water-related conflicts faced by households of different well-being groups. Water scarcity from drying water sources significantly affected households living on marginal uplands, who mostly rely on natural water sources for drinking water, sanitation, livestock feeding and small-scale irrigation. Results showed that about 44 percent of very poor households were still depending on traditional and natural sources of drinking water such as wells and streams, and over 87 percent of those households were facing water stress and conflicts related to water use. In many cases the public taps were far from the settlement of very poor households, as they were living in isolated marginal lands, or they were not able to afford maintenance fees, and thus depended upon the traditional and natural water sources.

The research findings (presented in greater detail in Gentle 2014) confirm that climate change impacts are experienced differentially. Although climate change vulnerability varies according to location of communities, occupation and gender of household head, overall the most significant factor influencing vulnerability is the well-being status of households within the communities (see Gentle et al. 2014). Major differences were due to higher sensitivity of very poor and poor households to food, water and health components, and lower adaptive capacity in terms of poor socio-demographic status, limited livelihood diversification strategies and a weak social network.

Access to information, services and resources; access to and control over natural resources such as land and water; and affordability of basic services such as water, health and credit remained major determining factors causing differential vulnerability. The findings are consistent with and provide further evidence to support previous research (Adger et al. 2003; Paavola and Adger 2002), which indicated that the poor within communities are differentially affected by the impacts of climate change. The finding has implications for the recently developed climate change policies and frameworks in Nepal (viz. Climate Change Policy 2011, NAPA 2010 and LAPA Framework 2011), which effectively ignore the notion of differential vulnerability in relation to vulnerability analysis and adaptation planning.

The findings contribute to a deeper understanding of the micro-level vulnerability of households and communities in local society and its causes, with evidence of the way in which the socio-economic context of households determine differential experiences of climate change impacts.

Capacity, commitment and willingness of CFUGs to support the most vulnerable for climate change adaptation in the context of differential impacts

The willingness of CFUGs to contribute to climate change adaptation, especially their willingness and commitment to reach the most vulnerable in the context of differential impacts of climate change, was assessed in terms of inclusion in

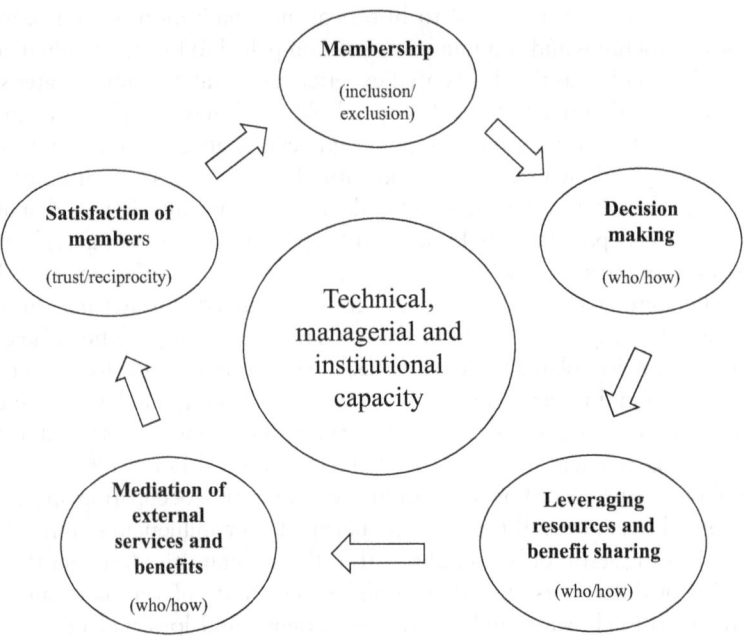

Figure 7.2 Framework to analyze capacity, willingness and success of CFUGs in climate change adaptation

membership, decision-making process, practices in leveraging resources and equity in benefit sharing, mediating external services and benefits, and satisfaction of user group members with decisions and outcomes (Figure 7.2).

Membership of CFUG

Membership of CFUGs is formal in nature as guided by forest policies of the government and constitutions of the user groups. One of the major forms of membership defined by the policies is based on traditional use rights. Almost all households residing close to the community forests were members of CFUGs. However, in many cases, recognition of traditional use rights was limited to certain caste groups who had claimed that a particular area of forest had been a private forest before nationalization of forests in 1957. For example, when Chisapani CFUG was handed over to the local community, membership was limited to certain upper caste groups, and Dalit households traditionally living in the village were excluded. Following a long debate, the forest office suggested that the users committee should include Dalits as members of the CFUG. However, the committee set a very high joining fee for Dalit households to become members. While some households paid the high fee to become members, others could not afford the high amount and were still excluded. The money collected as membership fees was then distributed amongst existing members (upper caste households) rather than depositing it in the CFUG account. The decision-making of this forest continues to be controlled by local elites and decisions are not transparent to other members.

Excluding poor and Dalit households from CFUG membership restricts their access to forest products, excludes them from participation in decision-making, and from other benefits arising from membership of the CFUG, which ultimately reduces capacity of poor and Dalits to adapt to climate change. Similar practices of exclusion of Dalit households from their membership exist in many CFUGs.

Decision-making

An analysis of Executive Committees of the four case study CFUGs revealed that the participation of well-off, medium, poor and very poor well-being groups in the committees in 2012 was 47 percent, 39 percent, 11 percent and 3 percent, respectively (Table 7.4). Out of 16 members in the key positions of Executive Committee, 56 percent were well-off and 44 percent were of medium wealth, whereas there was no representation by poor and very poor households in the key decision-making positions such as chairperson, vice chairperson, secretary and treasurer (Table 7.4).

The poor and very poor interviewees and FGD participants generally expressed the view that they were excluded from the decision-making process of CFUGs. In contrast, the well-off and elite members explained that the poor do not have the interest or capacity to contribute anything in the meetings and assemblies. While the complete lack of representation of the poorer households in decision-making positions is not the case in all CFUGs across Nepal, the opinions and examples presented above confirm that the decision-making of CFUGs is controlled by local elites based on their well-being status and traditional power. The decision-making authority of these CFUGs indicates who makes local rules (and thus holds power and influence), and how these rules affect the members, as seen in the following sections. As reported by poor members, exclusion of the poor from decision-making restricts their opportunity to raise their concerns and demands in response to CFUG rules and regulations, and thus the decisions of the CFUG often do not reflect their needs.

Table 7.4 Participation of different well-being groups in the decision-making positions of CFUGs

Well-being status	Total member households of CFUGs	Committee members	Key positions (chairperson, vice chairperson, secretary and treasurer) in the committee
Well-off	122 (30%)	18 (47%)	9 (56%)
Medium	155 (39%)	15 (39%)	7 (44%)
Poor	82 (20%)	4 (11%)	0
Very poor	43 (11%)	1 (3%)	0
Total numbers	402	38	16

(Source: CFUG records)

Leveraging resources and benefit sharing

The CFUGs collect revenue from member fees and fines, sale of forest products, and from donations and external support, and spend their funds on forest management, community development, income generation and livelihood improvement activities. The community forestry guidelines 2009 have a mandatory provision to spend at least 35 percent of CFUG income in pro-poor livelihood improvement activities. However, the financial records and audit reports of Kataharbari and Raniswanra CFUGs showed that from 2004 to 2011 the CFUGs had spent less than 5 percent of their income on pro-poor livelihood related activities. The CFUGs were investing their resources in local development and adaptation activities such as landslide control, soil conservation and water source protection on the public lands.

The implementation of pro-poor provisions as outlined in the revised community forestry guidelines is crucial to increase the adaptive capacity of poor and vulnerable communities. The guidelines include mandatory provisions to be carried out by all CFUGs such as: (i) well-being ranking of CFUG households according to their relative well-being; (ii) preparation and implementation of livelihood improvement plan for poor households; (iii) allocation of part of the CF land for income generating activities; and (iv) allocation of at least 35 percent of CFUGs' income for pro-poor activities. However, an assessment of the implementation status of pro-poor provisions as per the revised CF guidelines revealed that the CFUGs were not following most of the provisions. Although the CFUGs had carried out well-being ranking and mentioned the outcomes of the process in their constitutions or forest operational plans, the ranking was not used in deciding benefit-sharing provisions according to pro-poor principles (Table 7.5).

Table 7.5 Implementation status of pro-poor provisions in revised community forestry guideline

Major provisions	Implementation status in different CFUGs			
	Kataharbari	*Raniswanra*	*Chisapani*	*Manasalu*
Well-being ranking of CFUG households	Yes	Yes	Yes	Yes
Preparation and implementation of livelihood implementation plan	No	No	No	No
Allocation of CF land for income generating activities of poor	No	No	No	No
Allocation of at least 35% of CFUG income for pro-poor activities	Nominal amount of interest-free loan allocated to poor	No	No	No
Declaration of in-kind and cash contribution for poor in the annual plan	No	No	No	No

(Source: analysis of interviews, FGDs and CFUG records)

The local elites who were in decision-making positions of CFUGs had their own understanding of poverty and the behaviours of poor households, arguing that the poor and Dalits were responsible for their own situation by drinking alcohol, being lazy and not giving adequate attention to their own situation or its improvement. These interviewees reported that the CFUGs have a limited role in improving the livelihoods of the poor. However, very poor and poor interviewees described discriminatory practices and exclusion as the major causes pushing them into poverty and sustaining injustice. The poor and Dalits accused CFUGs of obstructing the delivery of pro-poor benefits provided by external organizations. So, it seems that while the government adopts a progressive approach in terms of drafting CF guidelines with pro-poor provisions, the implementation of these policies within communities and monitoring of the outcomes has not been effective.

Mediation of external services and benefits

One of the major expected roles of CFUGs is to mediate external services for the climate adaptation needs of local communities. The interviewees reported that local adaptation needs are multi-dimensional in nature and only possible to achieve by coordination of services by many organizations. Interviewees who were members of CFUG Executive Committees reported that they were able to influence district-level government (such as the forest, agriculture, women's development and soil conservation offices) for cash and material support in activities such as landslide control, soil conservation and other community development activities. Many research participants in decision-making positions described that CFUGs were successful in influencing and mediating negotiations with external organizations to deliver support for the adaptation needs of the communities.

However, in contrast, very poor and poor research participants claimed that CFUGs were actually hindering them in accessing benefits and services provided by government and other organizations. One district-level government interviewee from the forest office stated that the local elites in key positions of CFUGs were limiting the access of government officials to poor households for delivery of pro-poor provisions. In contrast to the expected role of CFUGs in influencing external agencies in favour of the poor and vulnerable, the institutions were accused of creating obstacles to the delivery of advice and services to benefit the poor households.

Given the tendency of government and national and international aid organizations to implement their programs in rural areas through local institutions (such as CFUGs) as implementing partners, the above analysis identifies a challenge for external organizations who seek to engage with vulnerable communities and support their adaptation needs through the platforms of CFUGs.

Satisfaction with services and functions of CFUGs

The survey data revealed that about 75 percent of respondents were very satisfied or satisfied with the overall functions and current services of CFUGs. However,

Table 7.6 Respondent satisfaction with CFUG functions and services

Level of satisfaction	Percentage of responses according to well-being groups			
	Well-off (n=38)	*Medium (n=50)*	*Poor (n=29)*	*Very poor (n=16)*
Very satisfied	47	24	7	0
Satisfied	47	62	52	25
Neutral	3	6	17	25
Not satisfied	3	8	24	50
Very dissatisfied	0	0	0	0

(Source: analysis of household survey data)

50 percent of very poor and about 24 percent of poor respondents expressed their dissatisfaction. Pearson's chi-square test showed a significant relationship between satisfaction with the current roles and functions of CFUGs and well-being status ($p < 0.01$) (Table 7.6).

Interviewees indicated their satisfaction with the roles and functions of CFUGs was mainly due to the contribution of CFUGs in managing local resources (such as forest and water) and accessing external services for local benefits. As reported by these interviewees, the efforts are successful in promoting collective actions such as forest protection, plantation and watershed management activities to reduce climate change vulnerability and risks in the communities. The very poor and poor interviewees who expressed dissatisfaction believed the CFUGs were established by local elites and well-off people for their own interest, and that CFUGs provide no benefits for the poor. Very poor and poor interviewees expressed frustration with the behaviour and practices of CFUGs such as exclusion from membership, discrimination in benefit sharing and the obstacles created for poor and vulnerable groups to access benefits and services.

Despite these issues and concerns, limited efforts had been made to improve the governance and transformation of CFUGs in favour of poor and vulnerable communities. Most of the persons in decision-making positions had connections with political parties and such political connections were used to gain power and to continue impunity.

The actions were not found to be effective in reducing individual- and household-level vulnerabilities, especially in addressing the vulnerabilities of the most vulnerable population. The exclusion and discrimination of vulnerable communities from access to and utilization of resources has been described as an injustice and a violation of human rights by very poor and poor communities.

The results presented above revealed that the elites (from upper caste, well-off economic status, derived from traditional or feudal legacy, educated, and persons with political or bureaucratic connections) are using their social, economic and political power to control the decision-making of CFUGs. According to very poor and socially marginalized people, local elites formed institutions to exercise power and to get benefits from local resources, through support provided by external agencies. It has been demonstrated that being in decision-making positions of CFUGs provides the power to make local rules and provisions that control who to include and exclude in membership, and who should receive various forms of benefits as an outcome of CFUGs. This process has systematically excluded very poor and socially marginalized communities from membership and, if members, from access to benefits from CFUGs.

Capacity of CFUGs to promote climate change adaptation

The CFUGs had no formal institutional policy and specific plan to support communities in climate change adaptation. Some CFUGs, however, were developing their policies and functions in response to the needs of local communities in the context of climate change. The existing knowledge, capacity and experience of all CFUGs was very low in designing and implementing climate change-related activities, and in addressing the needs of local community for climate change adaptation initiatives. The role and mandate of CFUGs was not explicitly identified in written policies; however the leaders in the decision-making positions of the CFUGs expressed willingness to enhance their understanding of climate change and implement adaptation activities to reduce vulnerability.

Most of the research participants and respondents reported important roles and potential of CFUGs in climate change adaptation. Almost all research participants in the communities had an association with CFUGs. The CFUGs had well-established networks with locally governed rules, norms and values to mobilize local communities and manage common pool resources. Interviewees identified the major roles and functions of CFUGs to support local communities in climate change adaptation as: (i) managing common pool resources such as forests; (ii) awareness and capacity building of local communities; (iii) leveraging and mobilizing social, financial and natural assets as per the local needs; (iv) supporting communities in disaster and emergencies; and (v) approaching and influencing external institutions for adaptation needs. The research found that the CFUGs achieved some success in reducing collective risks and vulnerabilities at the community level by applying forest protection, fire and grazing control, and watershed management activities. However, there were very limited examples of reduced individual- and household-level risks and vulnerabilities amongst the most vulnerable.

Discussion

CFUGs form the most extensive network of local institutions across rural Nepal and engage with over 60 percent of the rural population, thus presenting tremendous potential to utilize existing social and human capital to mediate external resources to support the local community in its efforts to adapt to climate change. This opportunity is particularly important for the poor and marginalized in society who have been found to be most vulnerable. The findings support the argument that local institutions play a crucial role in supporting communities in local-level adaptation planning and implementation, as highlighted by Agrawal and Perrin (2008) and Ostrom (2010). However, the existing knowledge, capacity and experience of CFUGs was inadequate in comparison to the increasing needs and demands of local communities for climate change adaptation.

One of the key findings is that the control of many CFUGs is mostly captured by local elites, as indicated by various authors in similar contexts. As a result of elite dominance, there is mistrust between the most vulnerable and those in decision-making positions of CFUGs. The poor and most vulnerable in the community lack trust in and ownership of CFUGs, largely due to exclusion from membership and decision-making, discrimination in sharing benefits, and the obstruction and manipulation of pro-poor services and activities by local elites. CFUGs are responsible for managing public goods; however, various forms of exclusion and discrimination are reported in benefit sharing, especially in addressing the needs and concerns of the poor and the most vulnerable. The findings support previous studies (Adhikari and Di Falco 2009; Jones and Boyd 2011) that reveal inequity in participation and benefit sharing based on caste, class and gender as major barriers of local institutions managing commons in similar contexts.

In many cases, the elites in CFUGs are promoting and perpetuating caste-, gender- and class-based discrimination, as well as the exclusion of poor and marginalized members of the community from access to resources, services and the benefit-sharing system. Local elites in the case study communities are not in favour of delegating authority and power to poorer community members. Unequal power relations and the continued dependency of the poor are benefiting local elites and sustaining unequal power relationships.

One major role identified for local institutions in climate change adaptation is the mediation of external intervention in favour of local needs and priorities (Agrawal 2010; Agrawal and Perrin 2008). However, our research revealed that the elite-led CFUGs are using the institutional platform to gain and sustain their traditional power and feudal legacies by making rules in favour of themselves and by developing biased implementation strategies. The findings cast doubt on the expected role of CFUGs to mediate external resources and services in favour of the most vulnerable.

There are very limited examples of reduced individual- and household-level risks and vulnerabilities amongst the most vulnerable. The success of CFUGs in managing forests as common pool resources is expected to reduce vulnerabilities and enhance adaptive capacity of local communities. However, most of the contributions are targeted at reducing collective vulnerabilities of communities, with little focus on addressing individual vulnerabilities, particularly amongst the most vulnerable. The study findings are in agreement with those of Adger (1999, 2003) from Vietnam, highlighting that individual and collective vulnerabilities were caused by different factors, and thus different strategies are needed to address these different vulnerabilities. Thus the effectiveness of adaptation strategies and responses should be evaluated in terms of their effectiveness in addressing individual, household and collective vulnerabilities.

The elite-dominated CFUGs are not in favour of prioritizing plans and allocating resources to address individual- and household-level adaptation needs of the most vulnerable. For example, CFUGs are spending most of their funds in local infrastructure development activities such as road construction, water source protection and tree plantation. However, resource allocation for pro-poor targeted activities is almost negligible, despite the mandatory policy provisions in the community forestry guidelines. While numerous authors have promoted the understanding that in managing commons, local institutions can develop their own institutional arrangements to manage local resources and to distribute benefits in an equitable and sustainable way (e.g. Andersson and Agrawal 2011; Ostrom 1992), these findings suggest that such outcomes are not always achieved. Indeed, the findings showed that local institutions can act against the sustainable and equitable distribution of benefits, so the existence of local institutions alone does not necessarily lead to such outcomes.

CFUGs, as local institutions, reflect a society, and function as a sub-set of that society. The discriminatory, exclusive and dominant practices adopted by the elites in decision-making positions of CFUGs may not have originated from within the CFUGs. Rather the practices may have been transferred from the society where such practices exist as a legacy of feudal, patriarchal and caste-based domination (Bennett et al. 2006). However, it is not clear whether civil society institutions (CFUGs in this case) may influence the society at large to reduce and end such anomalous practices, or whether those practices of CFUGs may only be reduced through societal influence.

The findings contribute to an understanding of how CFUGs may support local communities to reduce individual and collective risks and vulnerabilities. It is clear that the CFUGs, dominated by local elites, may not support the poor and most vulnerable households in reducing their individual- and household-level risks and vulnerabilities related to climate change. Despite these challenges and realities, service providers and external organizations still rely on CFUGs as implementing partners to deliver their services. The reliance on local institutions, such as CFUGs, without analyzing their internal governance, willingness and commitment towards the poor and most vulnerable may be inconsistent with the policies and objectives of service providers, and may work against their objectives of reaching the most vulnerable communities. Continuation of such partnerships without improving and transforming the structure and governance of CFUGs may further disenfranchise poor and vulnerable communities from the benefits of climate change adaptation initiatives.

Conclusion

Research has shown that CFUGs were well established in the Middle Hills of Nepal and relatively successful in managing local resources as common property and in reducing collective vulnerability in the context of climate change. Involvement of CFUGs in designing, implementing and coordinating climate change adaptation-related activities is an additional advantage because of their mandate to manage resources through a participatory process. CFUGs are considered by some external agencies to be the most appropriate entry point to deliver pro-poor initiatives to the poor in local communities, and hence become partners in adaption programs. Analysis of institutional capacity and governance of CFUGs from the perspectives of the most vulnerable was found to be an important consideration when informing policy and practice. Climate change adaptation in rural and remote hills of Nepal is occurring, and the roles, responsibilities and challenges of CFUGs as local institutions are expected to change to meet the expectations. Enhancing understanding, knowledge and skills on impacts of climate vulnerabilities as well as selection and implementation of appropriate adaptation measures are areas of improvement for CFUGs. Transformation in internal governance and the attitudes of decision-makers in CFUGs is required to equitably equip the most vulnerable communities to enable them in climate change adaptation. Transformation is also required at the agency and government levels in the implementation and monitoring of outcomes of policies and programs to enhance adaptation amongst the most vulnerable. The transformation should ultimately contribute in ensuring equity, efficiency and sustainability of the commons to address differential impacts of climate change.

Note

1 Schematic reproduction of map prepared by Kataharbari CFUG members during field-work in January 2012. The map showed that there is a 'power centre' of the village around the motor road at the valley floor where most of the well-off and politically influential people were living. Most of the services (such as high school, sub health post, cooperatives and shops) were located around the power centre. The power centre was surrounded by irrigated and fertile land. Some places of the village were identified as 'poverty pockets', mostly in the upland hills with non-irrigated lands, including some abandoned agriculture lands of those who have migrated from the village. The poverty pockets contained the residences of poor, Dalit, landless and powerless households. The poverty pockets were prone to landslides and drought and the households living in these areas were facing water stress.

References

Acharya, K.P. and Gentle, P. 2006, *Improving the effectiveness of collective action: Sharing experiences from community forestry in Nepal*, CAPRi Working Paper No. 54, pp. 17–21, July 2006, viewed 22 February 2017, http://ebrary.ifpri.org/cdm/ref/collection/p15738coll2/id/32951

Adger, W.N. 1999, Social vulnerability to climate change and extremes in coastal Vietnam. *World Development*, vol. 27, no. 2, pp. 249–269.

Adger, W.N. 2000, Institutional adaptation to environmental risk under the transition in Vietnam. *Annals of the Association of American Geographers*, vol. 90, no. 4, pp. 738–758.

Adger, W.N. 2003, Social capital, collective action, and adaptation to climate change. *Economic Geography*, vol. 79, no. 4, pp. 387–404.

Adger, W.N., Huq, S., Brown, K., Conway, D. and Hulme, M. 2003, Adaptation to climate change in the developing world, *Progress in Development Studies*, vol. 3, no. 3, pp. 179–195.

Adhikari, B. and Di Falco, S. 2009, Social inequality, local leadership and collective action: An empirical study of forest commons, *European Journal of Development Research*, vol. 21, no. 2, pp. 179–194.

Agrawal, A. 2010, Local institutions and adaptation to climate change, in R. Mearns and A. Norton (eds), *Social Dimensions of Climate Change: Equity and Vulnerability in a Warming World*, The World Bank, Washington, DC, pp. 173–198.

Agrawal, A. and Perrin, N. 2008, Climate adaptation, local institutions and rural livelihoods, *IFRI Working Paper # W081–6*, International Forestry Resources and Institutions Program, University of Michigan, Michigan, pp. 1–17, May 2008, viewed 22 February 2017, www.umich.edu/~ifri/Publications/W08I6%20Arun%20Agrawal%20 and%20Nicolas%20Perrin.pdf

Andersson, K. and Agrawal, A. 2011, Inequalities, institutions, and forest commons, *Global Environmental Change*, vol. 21, pp. 866–875.

Bennett, L., Tamang, S., Onta, P. and Thapa, M. 2006, *Unequal Citizens: Gender, Caste and Ethnic Exclusion in Nepal*, Department for International Development and The World Bank, Kathmandu, Nepal.

Daze, A., Ambrose, K. and Ehrhart, C. 2009, *Climate Vulnerability and Capacity Analysis (CVCA) Handbook*, Care International, viewed 9 October 2017, http://careclimat-echange.org/tool-kits/cvca/

DoF 2011, *CFUG database*, September 2011, Department of Forests, viewed February 2012.

DoF 2017, *CFUG database detail*, January 2017, Department of Forests, viewed 8 May 2017, http://dof.gov.np/image/data/Community_Forestry/Detail%20FUG%20All.pdf

Eakin, H. 2005, Institutional change, climate risk, and rural vulnerability: Cases from Central Mexico, *World Development*, vol. 33, no. 11, pp. 1923–1938.

Ebi, K.L., Woodruff, R., von Hildebrand, A. and Corvalan, C. 2007, Climate change-related health impacts in the Hindu Kush-Himalayas, *EcoHealth*, vol. 4, no. 3, pp. 264–270.

Gentle, P. 2014, *Equipping poor people for climate change: Local institutions and pro-poor adaptation for rural communities in Nepal*, PhD thesis, Charles Sturt University, viewed 22 February 2017, http://primo.unilinc.edu.au/primo_library/libweb/action/dlDisplay. do?vid=CSU2&docId=dtl_csu59617

Gentle, P. and Maraseni, T.N. 2012, Climate change, poverty and livelihoods: Adaptation practices by rural mountain communities in Nepal, *Environmental Science & Policy*, vol. 21, pp. 24–34.

Gentle, P., Thwaites, R., Race, D. and Alexander, K. 2014, Differential impacts of climate change on communities in the middle hills of Nepal, *Natural Hazards*, vol. 74, no. 2, pp. 815–836.

GoN 2009, *Guidelines for Community Forestry Program*, Ministry of Forests and Soil Conservation, Government of Nepal, Kathmandu, Nepal.

Jones, L. and Boyd, E. 2011, Exploring social barriers to adaptation: Insights from western Nepal, *Global Environmental Change*, vol. 21, no. 4, pp. 1262–1274.

Kanel, R.K. 2004, Twenty five years of community forestry: Contribution to millennium development goals, in R.K. Kanel, P. Mathema, B.R. Kandel, D.R. Niraula, A.R. Sharma and M. Gautam (eds), *Twenty-Five Years of Community Forestry*, Proceedings of the Fourth National Workshop on Community Forestry 4–6 August, 2004, Kathmandu, Nepal, pp. 4–18.

MoE 2010, *National Adaptation Programme of Action (NAPA)*, Ministry of Environment, Government of Nepal, Kathmandu.

Moser, C. 1996, *Confronting Crisis: A Comparative Study of Household Responses to Poverty and Vulnerability in Four Poor Urban Communities*, The World Bank, Washington, DC.

Mosse, D. 1994, Authority, gender and knowledge: Theoretical reflections on the practice of participatory rural appraisal, *Development and Change*, vol. 25, no. 3, pp. 497–526.

Narayan-Parker, D. 1997, *Voices of the Poor: Poverty and Social Capital in Tanzania*, The World Bank, Washington, DC.

Ostrom, E. 1992, *Crafting Institutions for Self-Governing Irrigation Systems*, Institute for Contemporary Studies, San Francisco, CA.

Ostrom, E. 2010, Polycentric systems for coping with collective action and global environmental change, *Global Environmental Change*, vol. 20, pp. 550–557.

Paavola, J., and Adger, W.N. 2002, *Justice and adaptation to climate change*, Tyndall Centre Working Paper No. 23, Tyndall Centre for Climate Change Research, October 2002.

Pokharel, B.K. and Byrne, S. 2009, *Climate change mitigation and adaptation strategies in Nepal's forest sector: How can rural communities benefit?* NSCFP Discussion Paper No 7, Nepal Swiss Community Forestry Project.

Richards, M., Maharjan, M. and Kanel, K. 2003, Economics, poverty and transparency: Measuring equity in forest user groups, *Journal of Forest and Livelihood*, vol. 3, no. 1, pp. 91–104.

Robledo, C., Fischler, M. and Patiño, A. 2004, Increasing the resilience of hillside communities in Bolivia, *Mountain Research and Development*, vol. 24, no. 1, pp. 14–18.

8 REDD+ and community forestry in Nepal

Mohan Poudel, Eak Rana and Richard Thwaites

Introduction

The previous chapter discussed the impacts of climate change in rural Nepal and the role of community forestry in contributing to climate change adaptation. This chapter explores efforts to mitigate climate change impacts through implementation of the global REDD+ policy, and the implications for community forestry (CF) and local communities in rural Nepal.

Under the United Nations Framework Convention on Climate Change (UNFCCC), a number of pilot REDD+ projects have trialled local implementation mechanisms, attracting considerable interest within government agencies, local communities and research institutions. In less developed countries, community forestry has often been seen by governments as key to implementing REDD+, though this perspective has also been challenged with arguments that REDD+ will likely undermine some of the critical strengths of CF.

This chapter explores the implementation of a REDD+ pilot in Nepal and the implications of this initiative for the local implementation institutions of community forestry. First, it highlights the challenges and opportunities for REDD+ considering the experiences of communities that have been involved to date. It then outlines the likely impacts of the implementation of REDD+ over the top of existing institutions, governance structures and forest management objectives. While the chapter presents a critical perspective of the potential positive and negative impacts of REDD+ on community forestry, it particularly seeks to identify possible opportunities and challenges for generating synergies between livelihood, environmental and carbon storage agendas.

Forests and emissions dynamics

Forests play a crucial role in regulating the world's climatic systems by maintaining the global carbon cycle (World Bank 2008), acting as both sink and source of carbon: photosynthesis turns atmospheric carbon into biomass and sugars, while respiration burns up some of these sugars, returning carbon back to the atmosphere.

However, forests and their capacity to deliver multiple services are at risk. In order to meet ever-increasing human needs, forests have been exploited and transformed since historical time. Forests are cleared, degraded and fragmented by timber harvest, conversion to agriculture, road building, human-caused fire, and in numerous other ways (Stern 2006). The Global Forest Resource Assessment (FAO 2011, p. 3) reveals that the world's forests have been decreasing at the rate of 5.2 million hectares per year since 2000. The loss and depletion of forests is a major issue for climate change (IPCC 2007, p. 36) with deforestation and forest degradation responsible for about 18 percent of global greenhouse gas emissions.

Overview of policy development and emerging concerns

Faced with the challenge of climate change and ongoing destruction of forests, the global community has collaborated in development of the 'Reducing Emissions from Deforestation and Forest Degradation and enhancement of carbon stocks' (or REDD+) program. REDD+ is an ambitious policy mechanism that aims to address drivers of deforestation and forest degradation and enhance carbon stocks in forests, particularly in tropical countries. The REDD+ policy mechanism has been unfolding since 2003, when the 9th Conference of Parties (COP9) of UNFCCC introduced 'compensated reduction' to account for deforestation. At COP13 the 'Bali road map' proposed a REDD mechanism to compensate developing countries and their communities for their forest conservation and regeneration efforts. The Poznan meeting in 2008 (COP14) acknowledged the role of conservation and sustainable management of forests in enhancement of forest carbon stocks, officially expanding REDD into REDD+. Box 8.1 highlights a chronology of REDD+ key initiatives in the UNFCCC.

Box 8.1   Chronology of REDD+ key initiatives

1997: COP3: Kyoto Protocol (KP) developed Clean Development Mechanism (CDM) to trade carbon emission offsets produced in developing countries

2003: COP9: Introduced 'compensated reduction' as approach to account for deforestation

2005: COP12: Papua New Guinea and Costa Rica submitted country position paper on REDD

2007: COP13: Adopted REDD as new forest and climate change regime; as successor of KP after 2012 – also known as 'Bali road map'

2008: COP14: REDD changed into REDD+, defined as reducing emissions from deforestation and forest degradation in developing countries. Piloting started

---

2009: COP15: Recognized REDD+ framework as a key to reducing global carbon emissions but no agreement was reached

2010: COP16: Agreed on a framework for REDD+ activities in developing countries

2011: COP17: Parties agreed to ensure environmental and social safeguards in their REDD+ policies; committed to respect indigenous knowledge and skills

2012: COP18: Extended KP to 2020 and expanded scope of REDD+ from forestry sector to other natural resource management sectors, including agriculture

2013: COP19: Continued REDD+ policy development, including rules for establishing reference levels, recognizing mitigation activities, creating institutions, ensuring safeguards and, above all, creating performance-based financing mechanisms

2014: COP20: Decisions were not explicit on REDD+, but both developed and developing countries agreed to set the targets of emission reductions with 'Intended Nationally Determined Contributions'

2015: COP21: Paris agreement reinforced objective to operationalize REDD+ as a result-based payment mechanism

As the REDD+ policy has unfolded, there is growing international consensus that REDD+ could be effective in reducing greenhouse gas emissions from the forestry sector by providing payment to developing countries to protect and better manage their forest resources. In addition, REDD+ could also yield additional social and environmental benefits (i.e. co-benefits), over the short and long term (Chhatre et al. 2012; Lee et al. 2011). Environmental co-benefits may include improved forest condition, conservation of biodiversity and protection of non-carbon forest ecosystem services. Expected social co-benefits may include livelihood improvements and benefits related to policy changes for implementation of REDD+ such as land tenure, improved governance, participation and decision-making processes.

While the main objective of REDD+ is to mitigate climate change by reducing net emissions and enhancing carbon stocks, the global policy of REDD+ could change management objectives of local forest management groups. Changes in objectives may raise some conflicts with existing management approaches where lives and livelihoods depend on access to forest resources. Emerging concerns are also related to customary access, use and tenure rights. Scholars (e.g. Dokken et al. 2014; Larson et al. 2013) have argued that REDD+ may overlook local communities' rights to access and use forest resources they have been managing.

REDD+ in Nepal

Nepal's engagement with REDD+ began when it became a member of the World Bank's Forest Carbon Partnership Facility (FCPF) scheme and submitted its

REDD Readiness Project Idea Note (R-PIN) to the World Bank in early 2008. In October 2009, Nepal joined the UN REDD program as an observer country and has since prepared a strategy for REDD+ implementation.

Nepal has made considerable progress in setting up an institutional framework for REDD readiness. The Ministry of Forests and Soil Conservation (MFSC) created a three-tiered framework consisting of the REDD Implementation Center (REDD IC); a multi-stakeholder REDD Working Group (RWG) with representation from government, experts, donors, and civil society organizations (CSOs); and a high-level, inter-ministerial REDD Apex Body. The REDD IC is responsible for coordinating the REDD+ readiness process under FCPF, as well as other REDD+ project initiatives in Nepal. The RWG and Apex Body are the official forums for REDD+ policy development and approval.

In addition to governmental efforts, some international organizations and local CSOs have undertaken REDD+ readiness initiatives, including piloting activities to explore the social and technical viability of REDD+ at the national and sub-national levels. These pilot projects were not designed to provide a full REDD+ model reflecting a performance-based approach required by carbon markets; rather, they have focused on social welfare activities with the aim to develop a REDD+ payment mechanism and demonstrate the effectiveness of benefit sharing. Current REDD+ initiatives in Nepal have mostly focused on three aspects of readiness processes: capacity building and awareness raising, development of benefit-sharing mechanisms, and measurement and monitoring of forest carbon stocks, while only a few REDD+ pilot initiatives (i.e. NORAD-funded REDD+ pilot) have distributed REDD+ incentive funds to local communities for their contribution to forest conservation. Given the experiences from preliminary REDD+ initiatives, Nepal is preparing to pilot a performance-based carbon trading scheme under the FCPF carbon fund support. The proposed sub-national Emission Reduction Program (ERP) covers 12 districts of the Terai Arc Landscape and is expected to start from early 2018.

In order to streamline REDD+ initiatives across the country, the Government of Nepal has recently developed Nepal's REDD+ strategy (REDD IC 2016b). This strategy proposes a 'nested' approach for REDD+ implementation whereby sub-national projects are implemented under a national policy framework. Such projects could then scale up to a national level as they strengthen capacity, improve governance and attain positive outcomes.

While REDD+ has been initiated through several forest management regimes in Nepal, the implementation of REDD+ through community forestry may have critical importance since CF provides a source of subsistence for local people and conservation of forest for both biodiversity and carbon emissions. Existing experiences of implementation of REDD+ and similar carbon mechanisms through community-managed forests have demonstrated both complementarities and conflicts between CF approaches and REDD+, resulting in both negative and positive outcomes for carbon stocks, local livelihoods, non-carbon forest ecosystem services and forest biodiversity (Alcorn 2014). We now move on to consider these complementarities and conflicts.

Community forestry and REDD+: can they go together?

Previous chapters have described CF as a mainstream forest management approach aiming simultaneously to address livelihood and conservation issues. The overall objectives of REDD+ schemes have been seen by many to complement those of the management of community forests such as to conserve forests and to support livelihood improvement (Agrawal and Angelsen 2009; Newton et al. 2016; Pelletier, Gélinas and Skutsch 2016).

While some aspects are common for both REDD+ and CF, there has been speculation that CF and REDD+ are incompatible in some aspects such as institutional arrangements, and priorities of REDD+ may conflict with the original objectives of CF (Bayrak and Marafa 2016; Mbow et al. 2012; Mustalahti et al. 2012). Such differences between CF and REDD+ reveal that implementing them together is not straightforward. Table 8.1 compares CF and REDD+ policy

Table 8.1 Complementarities and differences between CF and REDD+

Basis of comparison	Community forestry	REDD+ policy
Philosophy/ principles	CF's basic concepts and principles are based on the philosophy that people should participate in their own affairs. CF is founded on the assumption that local people are knowledgeable and capable regarding the environments in which they live, and their relationships to them, and that the active engagement of local people can enhance conservation outcomes	Economic incentives motivate developing countries/local communities to conserve forests and thereby reduce emissions. Problem of deforestation and forest degradation can be effectively tackled by incorporating knowledge of climate and the need for improvement of livelihoods and biodiversity
Priorities and objectives	Conservation of forest linking with improvement of local livelihoods and needs of local people. Encourage local communities to conserve and manage forest resources in their vicinity. Supply forest products and services (such as fuelwood, fibre and fodder) essential for the rural household and community level in a sustainable manner Food production and the environmental stability necessary for continued food production Generation of income and employment opportunity supporting local livelihoods	Conservation of forest linked with conservation and enhancement of carbon stocks and reducing net emissions. Financially reward developing countries and communities for their verified efforts to reduce emissions and enhance removal of greenhouse gases through a variety of forest management options

(Continued)

Table 8.1 Continued

Basis of comparison	Community forestry	REDD+ policy
Policy framework and approach	National policy framework applied at the local level. Local people control and make decisions over the conservation and management of forest resources, which are based on local values, needs and interests	Global policy framework with some provisions for country/local adjustment
Institutional scale	Mostly local	Sub-national, national

(Source: GoN 1993; GoN 1995; DoF 2009; MFSC 2010; REDD IC 2016a)

mechanisms based on their principles, objectives, approaches and institutional scales.

The comparison shows some fundamental differences between the CF and REDD+ policy mechanisms. These differences arise from the different objectives of CF to provide local people with forest products, livelihoods and income through extraction of timber, fuelwood and fodder, compared to the prioritized objective of REDD+ to enhance carbon stock by modifying behaviour and management. Many authors have commented on the contradiction within CF to provide both forest protection and local livelihoods. Charnley and Poe (2007) ask the question "whether it is realistic to expect community forestry to help conserve forests and also produce social and economic benefits for forest peoples", arguing that trade-offs may be required. Gilmour (2016) even proposes that objectives may be mutually exclusive. The contradiction is compounded by the overlaying of REDD+ onto CF and addition of the objective to increase carbon stock in the forest. The increased complexity of management for both carbon and forest product extraction implies complex trade-off decisions and additional burdens on local communities and households. Thus, the implementation of REDD+ through CF is not straightforward. The following section explores the experience of implementation of pilot REDD+ activities through CF in Nepal, particularly as experienced by local forest users.

REDD+ effects on local approaches to CF

REDD+ is seen as a value-laden policy instrument that has potential to induce changes in existing forest management practices. Hence, as local people seek to meet REDD+ requirements, REDD+ likely changes existing forest management strategies mainly limiting forest product use, benefit sharing and on-the-ground decision-making processes (Bryan and Crossman 2013; Ojija 2015). Studies have shown that implementation of the REDD+ pilot through CF has both positive and negative effects on local approaches to CF.

Reporting on an extensive multi-country metadata study of CF and its potentials for REDD+, Chhatre and Agrawal (2009) have argued that although CF could potentially become important for REDD+ success, it is also possible that carbon storage could become more dominant in forest management than the rights and interests of indigenous peoples and local communities. In a review of 23 REDD+ initiatives across six countries, Sills et al. (2014) suggest that limiting access to forest resources on which many smallholders rely for their primary income source could place local livelihoods at risk unless alternatives are offered. Similarly, studies of REDD+ pilot projects in Nepal by Maraseni et al. (2014) and Poudel et al. (2014) have shown that while REDD+ has resulted in improved governance of Community Forest User Groups (CFUGs), negative impacts are also apparent for livelihoods of forest-dependent people with REDD+ imposed restrictions on forest product extraction. Mutabazi et al. (2014) in Tanzania observed a loss of customary rights of local people; while Bottazzi et al. (2014) in Bolivia found a destabilization of existing forest benefit sharing as local people need to meet conditional requirements for accessing REDD+ payments.

These examples highlight the need for further consideration of the global experience of implementing REDD+ through existing CF regimes and whether the apparent contradictions between objectives can be resolved. To explore how REDD+ influences local CF approaches, case studies from two of the three REDD+ pilot watersheds in Nepal are illustrated in case studies 1 and 2. Poudel (2014) undertook case study research in the Ludhikhola watershed (Gorkha District) to explore the effects of REDD+ on local approaches to CF and identify ways for REDD+ and CF to work together (see also Poudel et al. 2014, 2015). As part of a study on trade-offs and synergies between carbon, livelihoods and biodiversity in the Charnawati Watershed (Dolakha District), Rana (2016) investigated the effects of REDD+ on ecological outcomes of CF in relation to changes in local CF practices (see also Rana, Thwaites and Luck 2016).

Case study 1: REDD+ effects on group governance and livelihood improvement

The REDD+ pilot in Ludhikhola watershed was jointly implemented from 2009 to 2013 by the International Centre for Integrated Mountain Development (ICIMOD), the Asian Network for Sustainable Agriculture and Bio-resources (ANSAB) and the Federation of Community Forestry Users, Nepal (FECO-FUN), with financial support from Norwegian Agency for Development Cooperation (NORAD).

Based on the experience of three CFUGs sharing similar socio-economic and biophysical characteristics (two that were part of the REDD+ pilot project and one CFUG adjacent to the Ludhikhola Watershed and not involved in the pilot), Poudel et al. (2014) investigated the impacts of the REDD+ pilot on community forestry and on the livelihoods of local forest-dependent people, exploring the factors that influence these impacts. The predominantly qualitative social research gathered data through 38 in-depth interviews, six focus-group discussions, review of secondary data and direct observations, supplemented by quantitative data from 91 households (at least 20 from each CFUG).

The area was characterized by remoteness and poverty but also diversity in culture, ethnicity and natural resources. Most people were subsistence farmers with high dependency on forest resources. To sustain their livelihoods over the last 30 years, local people have engaged with forest resource management through 31 CFUGs of different sizes across the watershed incorporating 3800 households.

The main objectives of the Ludhikhola pilot project were to demonstrate local REDD+ benefit-sharing mechanisms, and develop a mechanism for distribution of national REDD+ payments. To meet these objectives, each CFUG received annual REDD+ payments from the project Forest Carbon Trust Fund based on four criteria (carbon enhancement – 40 percent of ranking; ethnic diversity – 15 percent; proportion of women, indigenous people and Dalit – 15 percent; and proportion of poor people – 20 percent) assessed by a district-level monitoring committee. These REDD+ funds enabled CFUGs to provide seed grants for different activities that directly or indirectly support emissions reduction including forest enhancement, capacity building, and income generation (ICIMOD 2010).

The case study showed both positive and negative effects of REDD+ on established CF management structures (locally managed to meet local needs) and governance patterns (participatory and decentralized). Changes were identified in both group management, incorporating effects of REDD+ on decision-making processes, benefit sharing and capacity building, as well as in forest management, encompassing forest protection, development and utilization of forest resources.

On the positive side, CFUGs involved in the REDD+ pilot had adopted more regular meetings and transparent record keeping, auditing and reporting systems, suggesting that REDD+ has strengthened institutional capacity. Under the REDD+, CFUGs had also changed a traditionally practiced "equal for all" approach to benefit sharing, instead offering priority to women and marginalized and poorer households for opportunities such as support for income generating activities, skills development training (such as Local Resource Persons), and subsidized distribution of Improved Cooking Stoves (ICSs). No such changes were observed in the non-pilot case study site and local explanations of these changes suggest that REDD+ has stimulated a more pro-poor and pro-women (equity-based) approach to group governance in community forestry (though, as we shall see, this more equity-based approach has resulted in very limited change in opportunity for most poor and marginalized people).

On the other hand, REDD+ has threatened the decentralized power of decision-making at a community level fundamental to the existing devolved rights of CFUGs by imposing externally developed terms and conditions. For example, CFUGs were not allowed to spend seed grant funds beyond those activities framed at the central level by the project's regulating guideline. Despite highlighting the need for free, prior and informed consent of concerned stakeholders in the project document, the REDD+ pilot implementation did not ensure that all participants were informed. Marginalized women were found to lack the most basic information about the project and the benefits it offered. The research also found that REDD+ ignored customary rules that had always been an important characteristic of CF in Nepal by restricting traditionally practiced customary rights to access and use forest resources.

The research found that REDD+ has enhanced the equitable distribution of CFUG funds targeting poor and socially marginalized households such as Dalit and women. However, the distribution of these funds to households to support income generating activities was of limited effectiveness because the amount distributed from REDD+ to the CFUGs, and from CFUGs down to individual households, was inadequate. The reluctance of existing CFUG Executive Committee members to allocate adequate funds for income generating activities, poor integration of REDD+ benefit-sharing practices (following Forest Carbon Trust Fund guidelines) with existing CFUG benefit-sharing practice, and inadequate support to develop capacity of poor households inhibit the establishment of appropriate income generating activities. Information from research respondents and available records of study groups revealed that between 9 and 11 percent of the total households of each CFUG received REDD+ funds. Each household received nearly NRs 5,000 with a condition that it be paid back after one year. This suggests that REDD+ has enhanced pro-poor benefit sharing, but that this in itself is not enough, and that ensuring the benefits available compensate adequately for providing support to build capacity in poor households, as well as improved integration of REDD+ benefit sharing with existing practices in CFUGs, are essential to contribute to livelihood improvements of poor households. REDD+ also needs to consider the fundamental right of local forest users to receive adequate compensation for loss of access to resources and livelihood opportunities.

While REDD+ has enhanced institutional capacity of CFUGs, it is also likely to increase implementation costs for CFUGs to meet REDD+ requirements. Regardless of the size of the forest or the amount of incentives received, all CFUGs face the same monitoring and reporting requirements, so the relative cost to small CFUGs was much greater. With high implementation costs, the proportion of REDD+ funds available for livelihood improvement and forest activities was small, and thus the effectiveness of REDD+ in supporting rural livelihoods and conservation of forest carbon was reduced. So, given that most community forestry in Nepal is quite small in scale, the effectiveness and cost-efficiency of the REDD+ program as implemented is questionable.

Case study 2: REDD+ generates trade-offs between carbon, biodiversity and local livelihoods

Current debates on REDD+ through community-managed forests are marked by considerable uncertainty in terms of the ability to generate multiple benefits for forest biodiversity and local livelihoods, while also reducing carbon emissions. A case study by Rana (2016) in Charnawati watershed of Dolakha District in Nepal presents the findings related to the relationships between carbon, biodiversity (plant species diversity, stem density) and local livelihoods (collection of fuelwood, fodder and timber) resulting from the implementation of REDD+. The case study was carried out in 19 community forests across the Charnawati watershed involved in the REDD+ pilot initiative from 2009. These community forests are managed by local people of diverse ethnicities and different socio-economic status, interests and needs, and forest products from community forests provide a critical source of resources for local livelihoods.

Based on forest inventory data and CFUG records for 2010 and 2013, the case study showed that forest carbon stocks increased in all 19 CFs between 2010 and 2013, while forest biodiversity attributes and extraction of forest resources such as fuelwood and fodder decreased under the REDD+ pilot (Rana, Thwaites and Luck 2016). The changes in forest biodiversity attributes varied according to the location of CFs (altitude), duration of forest management (maturity of CFUG) and size of forests. However, extraction of forest products varied depending on the types and values of forest products. For example, the results show that extraction of fuelwood and fodder decreased after the implementation of REDD+, while extraction of timber increased over the same period. Variations in increase in carbon stocks were generally associated with the combined effects of biophysical characteristics, such as size of forest and presence of plant species in forest, socio-economic circumstances and institutional arrangements.

The research showed that REDD+ has enhanced forest conservation efforts of forest users resulting in changes in forest activities. The findings reveal that CFUGs have tightened restrictions on extraction of forest products, revitalized conservation activities, expanded forest patrolling, regulated cattle grazing and introduced alternative energies for cooking. While restricting extraction of forest products (i.e. fuelwood and charcoal) has contributed to carbon stock enhancement, it has undermined the customary rights of access to forest resources and livelihoods of forest-dependent people. This suggests that REDD+ can present trade-offs between carbon enhancement and local livelihoods.

Together, these case studies described above show both positive and negative effects of the REDD+ pilot have been experienced across biophysical and social and institutional aspects, largely at a community level. The major positive social and institutional effects include improved forest governance in terms of management procedures, increased participation of women and Dalit in decision-making bodies and promotion of equitable sharing of both monetary and non-monetary benefits (e.g. pro-poor-focused income generating activities, capacity building and quota system in delivering REDD+ benefits). With respect to biophysical outcomes, REDD+ may increase carbon stocks as local management emphasizes forest protection and enhancement of carbon stocks. Negative effects of REDD+ include the undermining of customary rights of forest-dependent people and threats to livelihoods resulting from restrictions being imposed on extractions of forest products. The funds available from REDD+ are totally inadequate to provide realistic support for income generating and livelihood improvement activities. Only a small number of households received funding and the amounts were deemed to be too small to support the establishment of income generating activities. Furthermore, REDD+ has ignored the need to enhance the skills and capacities of poor households to operate alternative livelihood activities. Compensation was not provided for loss of livelihoods resulting from the changed management practices. Finally, the effectiveness of REDD+ in improving livelihoods is also hindered if REDD+ benefit-sharing practices are not integrated with existing benefit-sharing practices of CFUGs.

Possible implications of REDD+ for CF

Based on the experiences of CFUGs in the Gorkha and Dolakha Districts presented in the above case studies, and from other studies identified above in the section discussing interactions between community forestry and REDD+, we now consider the governance, ecological and livelihoods implications of REDD+ for community forestry.

Governance and institutional implications

Lessons from the case studies suggest that grassroots institutions and governance arrangements developed for CF may be inadequate to meet the objectives of REDD+. As argued by Gilmour (2016), overlaying explicit carbon sequestration objectives onto existing CF objectives will greatly increase the operational complexities arising from the need to meet competing local, national and global objectives. Others, such as Bryan and Crossman (2013), Khatri et al. (2012), Ojija (2015) and Sills et al. (2014), have argued that under REDD+, CF might need to adapt its institutional architecture to incorporate the needs and interests of diverse stakeholders beyond the boundary of local communities. The resultant modification to local institutions and their governance systems guided by the needs of externally developed policies may produce negative as well as positive outcomes for CF, as explored in the following subsections.

Social equity at the community level may be enhanced

The Cancun safeguard principles (UNFCCC 2012) are designed to mitigate the potential negative impacts of REDD+ implementation on local environments and communities, and thus highlight the potential to improve social equity in existing CF regimes. Case study 1 presented in this chapter shows that positive discrimination towards the poor, marginalized and women can be applied to sharing benefits of CF and REDD+, and ensure representation of these groups in capacity building activities and in decision-making. Pro-poor-focused activities, such as the distribution of IGA seed grants and subsidies for ICSs to poorer households, and subsidized timber prices for single women and poor households are examples of how REDD+ can enhance equity of benefit sharing. Experiences from different pilot projects as described in the literature (e.g. Hagen 2014; Newton et al. 2016; Poudel et al. 2014) have also indicated that REDD+ is likely to improve CF governance by focusing on equitable distribution of benefits. However, provision of inadequate funds granted to only a few selected households can jeopardize social equity and therefore forest management in community-managed forests where collective behaviour depending on trust (see Dahal, Adhikari and Thwaites Chapter 6 this volume) is an essential element of forest conservation.

Decentralization process may be reversed

The above case studies show how REDD+ bundles CFUGs under a national or sub-national level network and imposes network-wide rules on all CFUGs, and could thus threaten decentralized forest governance. A number of authors (e.g. Bryan and Crossman 2013; Ojija 2015; Fisher 2014; Gilmour 2016; Newton et al. 2016; Phelps, Webb and Agrawal 2010; Poudel 2014; Sills et al. 2014; Sikor et al. 2010) have also argued that REDD+ could pose a threat to decentralized forest governance, diminishing CF's contribution to local autonomy, empowerment, ownership and community development. National REDD+ policies and approaches are not independent of global policy frameworks, and local REDD+ plans and policies are guided by national policies and frameworks. Such a top-down approach to developing policies and plans is inconsistent with the bottom-up, locally based policy planning and decision-making approaches of CF, diminishing local autonomy in decision-making and decentralized governance that have been identified as critical contributors to the success of CF, and potentially threatening the continued existence of CF in its current form. Despite efforts to promote community involvement in REDD+, requirements that REDD+ funding be distributed through government channels and spent according to externally developed rules may undermine decentralization (Phelps, Webb and Agrawal 2010). Further, requirement for a formal carbon monitoring system (e.g. a Monitoring, Review and Verification system or MRV) may require adoption of a centralized system, further undermining local autonomy.

Customary rules and practices may be ignored

The above case studies from Nepal reveal that the REDD+ pilot has focused on forest protection, neglecting local people's customary rights to access forest resources which form the basis of their livelihoods. Considering the introduction of grazing restrictions, customized harvesting and limited access to charcoal burning in the REDD+ pilot CFUGs, but not in the non-pilot CFUG, we conclude that REDD+ is likely to change customarily managed community forests into carbon-focused community protection forests. While application of customary rules has always been an important characteristic of CF in Nepal (Fisher 2014; Maraseni et al. 2014), REDD+ has put CF at risk by overlooking such customary rights to access and use forest resources. Other studies (e.g. Khatri et al. 2012; Maraseni et al. 2014; Poudel et al. 2014) also found that the REDD+ pilot overlooked customary rights and resource use practices of local communities, suggesting that REDD+ may shift emphasis towards a centralized system. It is quite possible that as local communities lose their rights to manage forests according to customary practices, their motivation to remain involved in community forestry will be challenged, and thus the sustainability of the CF model comes into question.

Ecological implications

The primary objective of REDD+ is to increase carbon stock in forests by reducing emissions through reduced deforestation and forest degradation, and enhancing forest biomass through sustainable management of forests. These activities seek to improve forest protection and maintain ecological integrity, with implications for growing stock and the function of the ecological system. These objectives and activities of REDD+ would appear to complement the environmental outcomes delivered by CF of reduced deforestation and forest degradation discussed in Chapter 4. We will now draw on what we have learned from the case studies and from the literature to explore the implications of REDD+ for ecological outcomes of CF.

Forest growing stock may be increased

The findings from case study two (above) reveal that carbon stocks in case study community forests increased by 3.56 tonnes per hectare between 2009 and 2013 (Rana 2016) and the increment of carbon stocks was related to direct and indirect measures of REDD+ piloting. Direct measures include regulations and practices to tighten access to forests and forest resources, such as through controlling grazing in the forest and customizing harvesting operations. Indirect measures include minimizing fuelwood consumption by providing ICSs and biogas, introducing IGAs and building capacity, particularly in fighting forest fires. Other recent research (e.g. Phelps, Webb and Adams 2012; Maraseni et al. 2014) including from CIFOR's Global Comparative Study (Angelsen et al. 2012) found that protection efforts under REDD+ result in enhanced forest condition, biodiversity and carbon stock. Based on data from Brazil, Cameroon and Indonesia, Resosudarmo et al. (2012) reported that households overwhelmingly perceived REDD+ to be focused on forest protection and enhancement of carbon stocks. A likely implication for CF is that while forest carbon stock may increase, forest governance may shift towards a more technical approach to support the use of advanced technologies in the complex task of monitoring changes in the forest. Such technical demands may threaten the role of local communities in monitoring and managing their own forest, a 'carbonization' or 'technicalization' of forest governance as recognized from studies in India by Vijge (2015).

Carbon may be considered as an added value for forest protection

REDD+ has established carbon as an added value of forests. Albbers and Robinson (2013) and Peskett (2011) argue that REDD+ has established carbon as a new form of property for local communities, creating the opportunity for financial benefits and livelihood support. The case studies discussed earlier in this chapter have also shown carbon money has been delivered to local communities for their contribution to increase forest carbon stock. However, carbon value of forests can influence forest management practices. Given that the previous intent of CF has been to provide for the forest resource needs of the local community in a sustainable manner, it would appear that there is a real danger of the 'tail wagging the dog', as suggested by Gilmour (2016), as carbon storage objectives become dominant over the rights and interests of local communities.

Displacement of forest destruction may occur beyond the boundary of REDD+ project

Better forest and ecological condition as a result of REDD+ may be limited to within the boundary of REDD+ projects. The case studies described in this chapter found evidence of displacement of human activities from pilot sites to non-pilot sites nearby, implying that imposing changes on forest management through REDD+ may increase the risk of forest destruction beyond the boundary of the project. This finding supports the UNFCCC recommendation (UNFCCC 2013) that REDD+ should be applied at the national scale, with sub-national REDD+ projects considered as an interim strategy to build capacity in implementing REDD+ policy effectively. However, establishing REDD+ at a national scale to overcome the displacement issue could pose a particular problem for community-based management approaches such as community forestry in Nepal by further undermining local customary practices, local rights and decentralized governance of forest management.

Livelihood implications

In his global review of 40 years of community-based forestry, Gilmour (2016) noted the importance of better understanding the trade-offs and synergies between carbon storage and livelihoods. The case studies in this chapter identified that trade-offs are being made between carbon storage and other forest benefits. The following sections discuss the implications of REDD+ for local livelihoods derived from CF.

Increased capacity for local people to undertake forest monitoring

Participatory forest monitoring has been described as a priority action in REDD+ policy documents (UNFCCC decisions, UN REDD and FCPF publications, country REDD+ strategies). The Cancun safeguard principles recognize the need for local engagement in REDD+ policy planning, implementing, and MRV process, and highlighted the need for local-level capacity building (UNFCCC 2011). Nepal's REDD+ strategy highlights community-based monitoring, measuring and reporting as a priority need for successful implementation of REDD+ (REDD IC 2016b). In order to develop local capacity to undertake REDD+ activities as envisaged, Nepal's readiness process has identified capacity building as a priority activity and has undertaken several training workshops, provided monitoring and measuring tools and developed guidelines (REDD IC 2016b). The above case studies reveal that training was provided for local resource persons to undertake participatory monitoring, measuring and reporting of REDD+ activities, enhancing capacity for communities to participate in the REDD+ MRV system, and delivering an alternative source of income to those individuals.

Livelihood trade-offs

The REDD+ interventions appear to have delivered improved forest condition, but this has been achieved at the expense of local livelihoods as access to forest resources is restricted. In particular, REDD+ may put poor and marginalized households under stress because they lack access to alternative forest resources to meet their daily requirements, including firewood and fodder (e.g. private trees, biogas and capacity to buy from markets). The likely implication is that the poorer may be working harder or travelling further to access the forest products they need for their subsistence. This ultimately increases food insecurity, along with income and other social insecurities, and also risks carbon displacement (leakage) elsewhere in the vicinity. Despite consuming more resources, however, households with private farmland and trees are likely to cope with limited supply from community forests (Neupane and Shrestha 2012).

Livelihood benefits do not outweigh losses

One of the underlying assumptions of REDD+ is that forest users will choose to conserve their forest if the compensation paid is higher than they would have obtained from alternative forest uses (Gilmour 2016). However, if the intention is that local communities gain additional livelihood benefits without affecting customary practices, evidence from the above case study research shows that the REDD+ pilot in Nepal has not only restricted customary use rights, but also failed to compensate adequately for the livelihood losses resulting from its activities. Other studies in REDD+ piloting sites (AIPP and IWGIA 2012; Bastakoti and Davidsen 2014; Chhatre et al. 2012; Maraseni et al. 2014) indicate that livelihood losses outweigh the benefits offered by REDD+. Similarly, as shown in the case studies, Maraseni et al. (2014) and Poudel et al. (2014) have argued that the seed grants provided in REDD+ pilots in Nepal undervalued the true costs to communities. There has been no reliable indication that upcoming REDD+ payment amounts would be more than the amounts provided in the piloting phase, resulting in speculation in the literature (AIPP and IWGIA 2012; Bastakoti and Davidsen 2014; Chhatre et al. 2012; Fisher 2014) that REDD+ will hamper rural livelihoods, for poor and marginalized people in particular. Experiences from the case studies in the light of literature suggest the following implications of REDD+ for rural communities in Nepal:

i   The poorer will be affected more, as they are more vulnerable to livelihood stress;

ii   Traditional occupations such as goat herders and blacksmiths will be displaced, without proper arrangements for substitutes; and

iii   CFUGs, poor and marginalized groups in particular, will be frustrated to an extent that may undermine the achievements of CF so far and its future sustainability.

These implications raise human rights concerns associated with REDD+ activities. REDD+ related human rights issues are particularly conspicuous with regard to matters concerning access to land and forest resources, as well as procedural rights concerning participation in the design and implementation of REDD+ policies (Savaresi 2013). As reported in the case studies, community forests that have been managed primarily to meet livelihoods and cultural requirements are likely to be more restricted in access and use of forest resources. Such changes may have significant human rights consequences, disrupting customary and cultural rights and forest-based livelihoods. A failure to safeguard human rights associated with the use and access to forest resources by local communities (poor, women, IPs and marginalized in particular) will mean REDD+ success is unlikely.

Opportunities and risks arising from REDD+ for CF

This section synthesizes and summarizes potential opportunities and risks of REDD+ to the CF system in Nepal based on the case study findings and other related literature.

Opportunities

The case studies findings described in this chapter indicate that REDD+ can render ecological, institutional and governance, and financial benefits to local people despite negative risks. These findings are consistent with other published findings. Albbers and Robinson (2013) and Peskett (2011) argue that REDD+ has established carbon as a new form of property for local communities, creating the opportunity for potentially large financial benefits and livelihood support. For example, at the Doha Conference of UNFCCC in 2012, developed countries committed to mobilize USD100 billion by 2020 for addressing climatic issues through mitigation and adaptation policy mechanisms, REDD+ in particular (Maraseni et al. 2014). The Paris agreement in 2016 (COP21) reinforces the objective to operationalize REDD+ as a results-based payment mechanism. This suggests that if appropriately designed and financial benefits are properly channelled to poor communities, and if adequate funds are made available, REDD+ has the potential to induce transformational change in livelihoods through improved forest governance, including decentralization, institutional strengthening, capacity building, income generation, and equitable and sustainable development (Korhonen-Kurki et al. 2014). In addition to the generic opportunities highlighted above, REDD+ offers the following specific opportunities to Nepal's CF policy and the local communities managing the forests.

- **REDD+ can enhance environmental co-benefits of CF:** The conservation of forests as a consequence of REDD+ may enhance non-carbon forest ecosystem services, providing additional potential benefits to local people (e.g. fuelwood). Improvement of forest condition enhances both carbon and biomass, thus increasing availability of some forest products to local people. The above case studies have described that REDD+ pilot CFUGs in Nepal observed improved conservation of forests as local people promoted forest regeneration and restoration, expanded enrichment plantation and fostered efforts to reduce incidence of forest fire, thus increasing both carbon stocks and production of forest products. In addition, case study 1 also reported protection of water sources, upstream downstream linkages and soil conservation as additional co-benefits of the REDD+ pilot in rural Nepal.

- **REDD+ can enhance biodiversity in CF**: Different forest activities associated with REDD+ can have positive biodiversity outcomes in community forests. While case study 2 reports a drop in biodiversity based on two selected indicators of plant diversity and stem density (which may result from particular forest management practices), others have argued that restoration of degraded forests and application of sustainable forest management practices can contribute to conservation of forests, thereby improving biodiversity in forests in the long run. REDD+-induced conservation of community forests can enhance landscape connectivity and biological corridors between community forests and other land use systems (Poudel et al. 2014). Furthermore, Alexander et al. (2011) posit that long-term incentives through REDD+ can motivate local communities to avoid human causes of forest degradation such as forest fire and grazing, yielding biodiversity outcomes.

- **REDD+ can strengthen CFUG governance**: Improvement of forest governance is considered as a social co-benefit of REDD+. Cromberg, Duchelle and Rocha (2014) and Mant et al. (2013) have described that changes in policies and rules for the implementation of REDD+ through CF can generate positive social outcomes such as clarification of land tenure and greater participation in decision-making process. Anderson and Zerriffi (2014), Thompson, Baruah and Carr (2011), and Luttrell et al. (2012) suggest that REDD+ can improve national and forest governance as local CFUGs must comply with international standards and agreements. The findings of case study 1 indicate that participation of Dalit and women increased with the implementation of REDD+ and distribution of benefits. This indicates that long-term implementation of REDD+ and distribution of benefits under REDD+ can improve participation of local forest users resulting in strengthened forest governance, which eventually can improve security of access rights of local communities to forest resources and strengthened equity in decision-making, benefit sharing and institutional arrangements.

Risks

These opportunities are not necessarily easy to realize. Global challenges include on-the-ground implementation of REDD+, governance and tenure issues, scope and cost of emissions reduction, revenue distribution mechanisms, concerns of indigenous communities, generation of credits and international REDD+

funding mechanisms. The following points synthesize potential risks of REDD+ to CF in Nepal.

- **Financial burden associated with increase in implementation costs**: Most community forests are small in size, and thus face relatively high REDD+ implementation costs for low additionality (verified increase in carbon storage). Studies in REDD+ pilot projects in Nepal (Maraseni et al. 2014; Newton et al. 2016; Poudel et al. 2015) have concluded that REDD+ payments may not cover the losses that local communities, particularly poor and disadvantaged households, will have to bear. Rana (2016) has found an increase in operational costs of CFUGs with the implementation of REDD+ through CFUGs in Nepal, associated with the higher number of Executive Committee meetings required to deal with the expanded agenda. Operational and implementation costs are also reported to have increased as a result of the record keeping, monitoring and reporting systems imposed on CFUGs as requirements of the REDD+ scheme. Yet with increased operational and implementation costs for CFUGs, only a small amount of REDD+ funds were available to support livelihood alternatives and income generating activities. This raises the concern for how future REDD+ projects can meet the higher operational and implementation costs of CFUGs, and compensate for the opportunity costs to households of reduced benefits from the CFUG. And if compensation (or incentive) payments are inadequate to cover opportunity costs, interest in participating in future REDD+ projects might be reduced, particularly for smaller CFUGs. At a household level, particularly for poor and Dalit households, the imposition of REDD+ through community forestry could de-motivate CFUG members from involvement in community forestry, placing a question over the future viability of some CFUGs.
- **The government may encourage recentralizing of CF**: REDD+ may encourage central governments to recentralize forest management authority from local communities by imposing forest conservation requirements that local communities are unlikely to meet (Phelps, Webb and Agrawal 2010; Springate-Baginski and Wollenberg 2010). As reported in case study 1 above, CFUGs may need to be bundled under a cluster of CFUGs compromising their independent identity and rights to make decisions. Such a bundling would demand a new centralized institution and governance system including decision-making and benefit-sharing mechanisms. Such a large and centralized institutional framework for REDD+ could make benefit sharing more difficult to ensure equity and accountability in verifying and compensating for carbon gains across CFUGs. Although CF policy in Nepal recognizes the use rights of CFUGs, arrangements for PES and carbon-related rights are not specified. If the situation remains unchanged, the motivation of community members to engage in REDD+ and conserve forests may be reduced. The lack of security of land tenure might also contribute to concerns around recentralization of CFUG governance responsibilities. While the government assigns access, management, and use rights of forests to CFUGs for

five- or 10-year periods based on the requirement to operate according to a constitution and operational plan, the government retains land ownership rights, and thus ultimate control over community forests.

- **Risks of emissions displacement:** One potentially significant risk of REDD+ implementation in CF is emissions displacement. Case study findings show that under the REDD+ pilot, local CFUGs reduced harvest of fuelwood and fodder from their community forests, while their dependence on private forests increased. Although the findings have not confirmed displacement of fuelwood and fodder demand to government-managed forests, they reveal that local people can shift their forest product demand to other sources (e.g. private forests) as long as alternatives to these forest products (e.g. alternative energies) are not adequately provided through REDD+; hence displacement occurs. In addition, some authors (Ojha et al. 2009; Maraseni et al. 2014; Poudel et al. 2014) have argued that displacement of forest products harvest between CFUGs and between community forestry and other management regimes (i.e. government-managed forest) in Nepal presents a potential risk of emissions displacement.

Conclusion

This chapter has briefly introduced theory, policies and processes of REDD+ development, described similarities and differences (or contradictions) between REDD+ and CF, illustrated case studies from REDD+ pilot sites in Nepal, and analyzed opportunities, challenges and likely implications of REDD+ in CF outcomes. Based on the issues raised, we now offer proposals for ensuring REDD+ complements CF enabling the achievement of the objectives of both REDD+ and CF.

In order to respect customary practices and address context-specific issues, REDD+ policies and measures should be based on CF policies and measures that have been considered successful in delivering multiple ecological and socio-economic outcomes.

Existing CF institutions do not explicitly acknowledge carbon enhancement as an objective of forest conservation. CFUGs usually develop their forest management plans to cover short-term periods (i.e. five to 10 years) focusing mostly on conservation and harvesting of forest products to meet their daily requirements. Incorporation of carbon enhancement as an objective of CF may require the development of a longer term plan that ensures ongoing delivery of existing short-term outcomes such as harvest of forest products. We believe this is likely to be a more effective approach to REDD+ implementation than imposing a carbon enhancement plan as an externally imposed requirement.

Maintaining customary rights and local autonomy (decentralized governance) is critical to ensuring the continued commitment of local people to CF. Any approach to implement REDD+ in CF will need to (i) ensure that CFUGs' rights to make autonomous decisions are not overlooked (avoid bundling affects); (ii) ensure that customary rights to access and use forests and forest resources to meet subsistence needs of forest-dependent people are continued; (iii) provide alternative options when resource access and supply is reduced; (iv) ensure women, Dalit and disadvantaged groups have equitable access to decision-making; (v) build leadership capacity to deliver efficient services and establish good governance, including equity and fairness in every step of the REDD+ process, from planning to benefit sharing; (vi) provide sufficient payment to local forest users to compensate for lost or reduced access to forest resources; and (vii) ensure that national CF policy acknowledges REDD+ and guarantees carbon rights to CFUGs.

The case study findings suggest that carbon money is not likely to outweigh losses that REDD+ will bring with it. As long as REDD+ funding is based on global market prices of carbon, and these prices remain at low levels that do not account for the foregone losses of local communities, a question remains over the viability of CF with the implementation of REDD+. There is a need for a detailed and critical cost-benefits analysis of REDD+ activities to assess their influence on local people's livelihoods, well-being and potential contribution to the sustainable development of CF. In this case, development of a payment scheme for non-carbon ecosystem services (i.e. biodiversity conservation and water services) along with carbon through REDD+ could compensate for the contribution of local people to conservation, leading to enhanced long-term viability of CF.

Overall, if REDD+ is going to limit access to forest resources for a global good (i.e. carbon), then it can be argued that it is a fundamental right that the local users should receive adequate compensation for what they have lost. The seed grants made available in the Ludhikhola pilot project would appear to be totally inadequate and thus these local human rights are being denied. This raises the question for the future of CF as a locally based management approach. By imposing rules from outside challenging local autonomy, and by denying human rights to resources or adequate compensation for their loss, neither CF nor REDD+ are likely to have much of a future. The question also arises whether the global market for carbon can ever truly match the costs imposed on local forest users, for which REDD+ should be morally bound to meet.

References

Agrawal, A. and Angelsen, A. 2009, Using community forest management to achieve REDD+ goals, in A. Angelsen, M. Brockhaus, M. Kanninen, E. Sills, W.D. Sunderlin and S. Wertz-Kanounnikoff (eds), *Realising REDD+: National Strategy and Policy Options*, Vol. 1, CIFOR, Bogor, Indonesia, pp. 201–212.

AIPP and IWGIA 2012, *Briefing Paper on REDD+, Rights and Indigenous Peoples: Lessons From REDD+ Initiatives in Asia*, Asian Indigenous Peoples Pact (AIPP) and International Work Group from Indigenous Affairs (IWGIA), viewed 25 May 2017, www. iwgia.org/iwgia_files_publications_files/0655_Doha_briefing_Final_Artork.pdf

Albbers, H.J. and Robinson, E. 2013, Reducing emissions from deforestation and forest degradation, in J.F. Shogren (ed.), *Encyclopedia of Energy, Natural Resource, and Environmental Economics*, Elsevier, Amsterdam, Netherlands, pp. 78–85.

Alcorn, J.B. 2014, *Lesson Learned From Community Forestry in Latin America and Their Relevance for REDD+*, USAID, Washington, DC.

Alexander, S., Nelson, C.R., Aronson, J., Lamb, D., Cliquet, A., Erwin, K.L. and Higgs, E.S. 2011, Opportunities and challenges for ecological restoration within REDD+, *Restoration Ecology*, vol. 19, no. 6, pp. 683–689.

Anderson, E. and Zerriffi, H. 2014, *The effect of REDD+ on forest people in Africa: Access, distribution and participation in governance*, Responsive Forest Governance Initiative (RFGI) Working Paper No. 1, CODESRIA, Dakar, Senegal.

Angelsen, A., Brockhaus, M., Sunderlin, W.D. and Verchot, L.V. 2012, *Analyzing REDD+: Challenges and Choices*, CIFOR, Bogor, Indonesia.

Bastakoti, R.R. and Davidsen, C. 2014, REDD+ and forest tenure security: Concerns in Nepal's community forestry, *International Journal of Sustainable Development and World Ecology*, vol. 21, no. 2, pp. 168–180.

Bayrak, M.M. and Marafa, L.M. 2016, Ten years of REDD+: A critical review of the impact of REDD+ on forest-dependent communities, *Sustainability*, vol. 8, no. 7, art. 620.

Bottazzi, P., Crespo, D., Soria, H., Dao, H., Serrudo, M., Benavides, J.P., Schwarzer, S. and Rist, S. 2014, Carbon sequestration in community forests: Trade-offs, multiple outcomes and institutional diversity in the Bolivian Amazon, *Development and Change*, vol. 45, no. 1, pp. 105–131.

Bryan, B.A. and Crossman, N.D. 2013, Impact of multiple interacting financial incentives on land use change and the supply of ecosystem services, *Ecosystem Services*, vol. 4, pp. 60–72.

Charnley, S. and Poe, M.R. 2007, Community forestry in theory and practice: Where are we now? *Annual Review of Anthropology*, vol. 36, pp. 301–336.

Chhatre, A. and Agrawal, A. 2009, Trade-offs and synergies between carbon storage and livelihood benefits from forest commons, *Proceedings of the National Academy of Sciences*, vol. 106, no. 42, pp. 17667–17670.

Chhatre, A., Lakhanpal, S., Larson, A.M., Nelson, F., Ojha, H. and Rao, J. 2012, Social safeguards and co-benefits in REDD+: A review of the adjacent possible, *Current Opinion in Environmental Sustainability*, vol. 4, no. 6, pp. 654–660.

Cromberg, M., Duchelle, A.E. and Rocha, I.D.O. 2014, Local participation in REDD+: Lessons from the Eastern Brazilian Amazon, *Forests*, vol. 5, no. 4, pp. 579–598.

DoF 2009, *Operational Guideline for Community Forestry*, Department of Forests, Kathmandu, Nepal.

Dokken, T., Caplow, S., Angelsen, A. and Sunderlin, W.D. 2014, Tenure issues in REDD+ pilot project sites in Tanzania, *Forests*, vol. 5, no. 2, pp. 234–255.

FAO 2011, *State of the World's Forests 2011*, Food and Agriculture Organization of the United Nations, Rome, Italy.

Fisher, R.J. 2014, *Lesson Learned From Community Forestry in Asia and Their Relevance for REDD+*, USAID-supported Forest Carbon, Markets and Communities (FCMC) Program, Washington, DC.

Gilmour, D. 2016, *Forty Years of Community-Based Forestry: A Review of Its Extent and Effectiveness*, Food and Agriculture Organization of the United Nations, Rome, Italy.

GoN 1993, *Forest Act 1993*, Government of Nepal, Kathmandu.

GoN 1995, *Forest Regulation 1995*, Government of Nepal, Kathmandu.

Hagen, R. 2014, *Lessons Learned From Community Forestry and Their Relevance for REDD+*, USAID-supported Forest Carbon, Markets and Communities (FCMC) Program, Washington, DC.

ICIMOD 2010, *Operational Guidelines of Forest Carbon Trust Fund*, International Centre for Integrated Mountain Development (ICIMOD), Kathmandu, Nepal.

IPCC 2007b, Climate change 2007: Synthesis report, in R.K. Pachauri and A. Reisinger (eds), *Contribution of Working Groups I, II and III to the Fourth Assessment Report of the*

Intergovernmental Panel on Climate Change (Synthesis Report: SYR 6.3), p. 104, IPCC, Geneva, Switzerland.

Khatri, D.B., Paudel, N.S., Bista, R. and Bhandari, K. 2012, *Review of REDD+ Payment Mechanism Under Pilot Project: Implications for Future Carbon Payments in Nepal*, Forest Action Nepal, Kathmandu, Nepal.

Korhonen-Kurki, K., Sehringa, J., Brockhaus, M. and Gregorio, M.D. 2014, Enabling factors for establishing REDD+ in a context of weak governance, *Climate Policy*, vol. 14, no. 2, pp. 167–186.

Larson, A.M., Brockhaus, M., Sunderlin, W.D., Duchelle, A., Babon, A., Dokken, T., Pham, T.T., Resosudarmo, I.A.P., Selaya, G., Awono, A. and Huynh, T.B. 2013, Land tenure and REDD+: The good, the bad and the ugly, *Global Environmental Change*, vol. 23, no. 3, pp. 678–689.

Lee, D., Seifert-Granzin, J., Neeff, T., Göhler, D., Liss, B. and Busch, A. 2011, *Maximizing the co-benefits of REDD-plus actions*, Discussion Paper for a Regional Expert Workshop supported by the German International Climate Initiative, 27–29 September 2011, Subic, Philippines, viewed 25 May 2017, www.international-climate-initiative.com/fileadmin/Dokumente/ICI-Workshop_REDDplus-non-carbon_benefits-safeguards-discussion_paper_839.pdf

Luttrell, C., Loft, L., Gebara, F. and Kweka, D. 2012, Who should benefit and why? Discourses on REDD+ benefit sharing, in A. Angelsen (ed.), *Analysing REDD+: Challenges and Choices*, CIFOR, Bogor, Indonesia.

Mant, R., Swan, S., Anh, H.V., Phuong, V.T., Thanh, L.V., Son, V.T., Bertzky, M., Ravillous, C., Thorley, J., Trumper, K. and Miles, L. 2013, *Mapping the Potential for REDD+ to Deliver Biodiversity Conservation in Viet Nam: A Preliminary Analysis*, UNEP-WCMC, Cambridge, UK; SNV, Ho Chi Minh City, Vietnam.

Maraseni, T.N., Neupane, P.R., Lopez-Casero, F. and Cadman, T. 2014, An assessment of the impacts of the REDD+ pilot project on community forests user groups (CFUGs) and their community forests in Nepal, *Journal of Environmental Management*, vol. 136, pp. 37–46.

Mbow, C., Skole, D.L., Dieng, M., Justice, C., Kwesha, D., Mane, L., El Gamri, M., von Vordogbe, V. and Virji, H. 2012, *Challenges and prospects for REDD+ in Africa: Desk review of REDD+ implementation in Africa*, GLP Report No. 5, GLP-IPO, Copenhagen, Denmark.

MFSC 2010, *Nepal's Readiness Preparation Proposal (RPP) for REDD*, Ministry of Forests and Soil Conservation (MFSC), Government of Nepal, Kathmandu, Nepal.

Mustalahti, I., Bolin, A., Boyd, E. and Paavola, J. 2012, Can REDD+ reconcile local priorities and needs with global mitigation benefits? Lessons from Angai Forest, Tanzania, *Ecology & Society*, vol. 17, no. 1, art. 16.

Mutabazi, K.D., George, C.K., Dos Santos, A.S. and Felister, M.M. 2014, Livelihood implications of REDD+ and costs-benefits of agricultural intensification in Redd+ pilot area of Kilosa, Tanzania, *Journal of Ecosystem and Ecography*, vol. 4, no. 2, art. 144.

Neupane, S. and Shrestha, K. 2012, Sustainable forest governance in a changing climate: Impacts of REDD program on livelihoods of poor communities in Nepal's community forestry, *OIDA, International Journal of Sustainable Development*, vol. 4, no. 1, pp. 71–82.

Newton, P., Oldekop, J.A., Brodnig, G., Karna, B.K. and Agrawal, A. 2016, Carbon, biodiversity, and livelihoods in forest commons: Synergies, trade-offs, and implications for REDD+. *Environmental Research Letters*, vol. 11, no. 4, 044017.

Ojha, H.R., Dahal, N., Baral, J., Subedi, R. and Branney, P. 2009, *Making REDD functional in Nepal: Action points for capitalizing opportunities and addressing challenges*, Discussion

Paper (draft), Forestry Nepal, Kathmandu, viewed 9 August 2012, www.forestrynepal.org/publications/article/4043

Ojija, F. 2015, Assessment of current state and impact of REDD+ on livelihood of local people in Rungwe District, Tanzania, *Forest Research*, vol. 4, no. 4, art. 156.

Pelletier, J., Gélinas, N. and Skutsch, M. 2016, The place of community forest management in the REDD+ landscape, *Forests*, vol. 7, no. 8, art. 170.

Peskett, L. 2011, *Benefit Sharing in REDD+: Exploring the Implications for Poor and Vulnerable People*, World Bank, Washington, DC.

Phelps, J., Webb, E.L. and Adams, W.M. 2012, Biodiversity co-benefits of policies to reduce forest-carbon emissions, *Nature Climate Change*, vol. 2, no. 7, pp. 497–503.

Phelps, J., Webb, E.L. and Agrawal, A. 2010, Does REDD+ threaten to recentralize forest governance? *Science*, vol. 328, no. 5976, pp. 312–313.

Poudel, M.P. 2014, *Examining outcomes of REDD+ through community forestry in rural Nepal*, Doctoral dissertation, viewed 25 May 2017, http://primo.unilinc.edu.au/primo_library/libweb/action/dlDisplay.do?vid=CSU2&docId=dtl_csu75928

Poudel, M., Thwaites, R., Race, D. and Dahal, G.R. 2014, REDD+ and community forestry: Implications for local communities and forest management: A case study from Nepal, *International Forestry Review*, vol. 16, no. 1, pp. 39–54.

Poudel, M., Thwaites, R., Race, D. and Dahal, G.R. 2015, Social equity and livelihood implications of REDD+ in rural communities: A case study from Nepal, *International Journal of the Commons*, vol. 9, no. 1, pp. 177–208.

Rana, E. 2016, *REDD+ and ecosystem services trade-offs and synergies in community forests in Central Himalaya, Nepal*, Doctoral dissertation, viewed 25 May 2017, http://primo.unilinc.edu.au/primo_library/libweb/action/dlDisplay.do?vid=CSU2&docId=dtl_csu88032

Rana, E., Thwaites, R. and Luck, G. 2016, Trade-offs and synergies between carbon, forest biodiversity and forest products in Nepal community forests, *Environmental Conservation*, vol. 44, no. 1, pp. 5–13.

REDD, I.C. 2016a, *REDD+ Readiness Self Assessment Report (R-Package) Nepal*, REDD Implementation Center, Kathmandu, Nepal.

REDD, I.C. 2016b, *REDD+ Strategy Nepal* (draft), REDD Implementation Center, Kathmandu, Nepal.

Resosudarmo, I.A.P., Duchelle, A.E., Ekaputri, D. and Sunderlin, W. 2012, Local hopes and worries about REDD+ projects, in A. Angelsen (ed.), *Analyzing REDD+: Challenges and Choices*, CIFOR, Bogor, Indonesia.

Savaresi, A. 2013, REDD+ and human rights: Addressing synergies between international regimes, *Ecology and Society*, vol. 18, no. 3, pp. 5–13.

Sikor, T., Stahl, J., Enters, T., Ribot, J.C., Singh, N., Sunderlin, W.D. and Wollenberg, L. 2010, REDD-plus, forest people's rights and nested climate governance, *Global Environmental Change*, vol. 20, no. 3, pp. 423–425.

Sills, E.O., Atmadja, S.S., de Sassi, C., Duchelle, A.E., Kweka, D.L., Resosudarmo, I.A.P. and Sunderlin, W.D. (eds) 2014, *REDD+ on the Ground: A Case Book of Subnational Initiatives Across the Globe*, CIFOR, Bogor, Indonesia.

Springate-Baginski, O. and Wollenberg, E. (eds) 2010, *REDD, Forest Governance and Rural Livelihoods: The Emerging Agenda*, CIFOR, Bogor, Indonesia.

Stern, N. 2006, *The Economics of Climate Change: The Stern Review*, Cambridge University Press, Cambridge, UK.

Thompson, M.C., Baruah, M. and Carr, E.R. 2011, Seeing REDD+ as a project of environmental governance, *Environmental Science & Policy*, vol. 14, no. 2, pp. 100–110.

UNFCCC 2011, *Report of the conference of the parties serving as the meeting of the parties to the Kyoto Protocol on its seventh session, held in Durban from 28 November to 11 December 2011*, Decisions adopted by the conference of the parties (FCCC/KP/CMP/2011/10/Add.1), UNFCCC, Durban, viewed 25 May 2017, http://unfccc.int/resource/docs/2011/cmp7/eng/10a01.pdf

UNFCCC 2012, *Report of the conference of the parties on its eighteenth session, held in Doha from 26 November to 8 December 2012*, Decisions adopted by the conference of the parties (FCCC/CP/2012/8/Add.1), UNFCCC, Doha, viewed 25 May 2017, http://unfccc.int/resource/docs/2012/cop18/eng/08a01.pdf

UNFCCC 2013, *Report of the conference of the parties on its nineteenth session, held in Warsaw from 11 to 23 November 2013*, Decisions adopted by the conference of the parties (FCCC/CP/2013/10/Add.1), UNFCCC, Warsaw, viewed 25 May 2017, http://unfccc.int/resource/docs/2013/cop19/eng/10a01.pdf

Vijge, M.J. 2015, Competing discourses on REDD+: Global debates versus the first Indian REDD+ project, *Forest Policy and Economics*, vol. 56, pp. 38–47.

World Bank 2008, *Forests Source Book: Practical Guidance for Sustaining Forests in Development Cooperation*, The World Bank, Washington, DC.

9 Labour migration, the remittance economy and the changing context of community forestry in Nepal

Krishna Shrestha and Robert Fisher

Introduction

Community forestry (CF) in Nepal and elsewhere is facing unprecedented social and economic change. Initially, it emerged with the premise that people's meaningful role in decisions affecting surrounding forests can achieve improved socio-economic well-being and ecological sustainability, and also as a means to meet the basic needs of rural households (FAO 1978). CF ideals have had strong focus on local-level subsistence and protection of forests. However, this focus has shifted. CF has now been explicitly linked to livelihoods (Cedamon et al. 2017), food security (Khatri et al. 2016) and climate change (Ojha et al. 2013), and with the advent of globalization, urbanization and economic development, increased infrastructure, communication and household incomes, local communities in developing countries are increasingly involved in market economies (Agrawal, Chhatre and Hardin 2008; Paudel 2016; Gilmour 2016a; Ojha et al. 2017). This has brought considerable changes in the environment in which community forestry operates.

Nepal's community forestry program emerged in the 1970s and is now recognized as one of the world's most widely practiced systems of community-based natural resource management. It involves over 19,361 forest user groups managing 1,813,478 hectares of national forest (over one-quarter of Nepal's total forest area) and it has involved 2,461,549 households (DoF 2017). This is one of the priority programs of the Nepal government, as about 35 percent of the total development budget allocated to the forest ministry is spent on the Community Forestry Program (GoN 2013). Nepal's CF legislation is often lauded as a progressive example of CF in terms of legally recognizing the local control of forests (Ojha et al. 2007; Pokharel 2011; Ojha 2013; Sunam, Paudel and Paudel 2015). Initially the use of forests was for subsistence purposes such as for the provision of firewood and fodder for domestic livestock (Khatri et al. 2016), but now the forest is connected to markets, and local communities are in a position to sell forest products if there were less administrative and legislative barriers (Gilmour 2016b). Nepal's CF has also been used for experimenting with REDD+ (Ojha et al. 2015; Poudel, Rana and Thwaites, Chapter 8 this volume). Forests and CF are now seen as a source of prosperity through "sustainable forest management" (GoN 2013), with a national workshop on this matter having been organized by the Department of Forests in February 2017.

While CF has advanced considerably, Nepal has been going through unprecedented socio-economic and environmental changes in recent years. One notable change is the labour migration overseas from Nepal's villages and towns, and the resulting emergence of a remittance economy. This has had wide effects on Nepal, from effects on the national economy to local people's lives and livelihoods, including the way CF is seen and managed. About 4 million adult males (out of a population of about 28 million) have left Nepal to engage in wage labour in the Middle East and elsewhere. Remittances from Nepalese outside Nepal represent over 29.2 percent of Nepal's GDP – one of the highest proportions of remittances to GDP in the world (World Bank 2016). This figure was only 16 percent of GDP in 2006. While the national development discourse highlights the great gift of remittance to poverty reduction in Nepal, the impacts of the remittance economy have not been completely positive (Tiwari and Bhattarai 2011; Devkota 2014). While this point is not meant to underplay the contributions of the remittance economy in enhancing the income and livelihoods of some households, the negative consequences should not be underestimated. They are manifested in varied ways, such as in the form of feminization of agriculture, increasing gender injustices in farm activities, and negative impacts to family and social values, culture and practices (Paudel, Tamang and Shrestha 2014). With many adult males away from the villages, productive lands are under-utilized due mainly to labour shortage, leading to the decrease in food production and acute food insecurity in a country that has a chronic food insecurity problem (Tamang, Paudel and Shrestha 2014). Moreover, village feminization combined with rapid migration and an ongoing national political crisis has led to the collapse of local institutions and community-based decision-making practice (Paudel 2016; Shrestha and Ojha 2017; Ojha et al. 2017).

With such unprecedented changes associated with the remittance economy that has penetrated Nepal's rural villages and towns, local institutions such as Community Forest User Groups (CFUGs) and the management of environment through CF have been profoundly affected. The important question is how and to what extent such effects are felt, and what future lies ahead for CFUGs and sustainable management of community forests. There is no sufficient data or robust analysis of CF in the context of the remittance economy and this chapter aims to fill this gap. The principal purpose of the chapter is therefore to explore how the remittance economy has changed the environment within which CF operates, and whether and how the remittance economy helps (or hinders) the progress of CF. In particular, this chapter will seek to answer the following questions:

- How is CF situated in the broader context of Nepal, and how is CF likely to be affected by recent changes in Nepal's society, economy and politics?
- What are the consequences of global changes, notably with the advent of remittance economy, on the local management of CF? Does the remittance economy help (or hinder) CF?
- How can Nepal's CF be more resilient in the changing national and global contexts?

The chapter is structured in five sections. Following the introduction, the second section describes research methods employed in this study. The third section will elaborate on the emerging issues in Nepal's CF in the context of national economy, politics and society. In the fourth section, we will discuss the changing context of CF and the rise of the remittance economy, and demonstrate how the functioning and sustainability of CF is affected by the rise of the remittance economy. The fifth section will explain how CF can remain resilient in the changing national and global contexts. Finally, the chapter concludes by highlighting the fact that the remittance economy is a threat to the sustainability of CF. We argue that for CF to continue advancing in the new context of a neoliberal political economy, it needs to be rethought in such a way that its planning process is reframed and CF institutions consider the remittance economy as an opportunity, actively utilize forests for equitable development and see markets as a platform to help alleviate poverty.

Research methods

This chapter is based both on a broad literature review of the remittance economy and its impacts, and on an in-depth case study of two Community Forest User Groups (CFUGs) in Dhamilikuwa Village Development Committee (VDC) of Lamjung District in western Nepal. The case study provides information specifically on the issue of remittances and their social and economic impacts as well as the consequences to community forestry. In selecting this VDC, the researchers considered different criteria including the extent of labour out-migration in recent years, community heterogeneity and the state of CF functioning. The study was part of a research program[1] that had the objective of exploring changes in life and livelihoods of CFUGs in rural Nepal in response to national and global challenges. This paper draws on ethnographic field research conducted as part of this study, comprising four field visits between 2013 and 2016. In 2013, the first author of this chapter started the research process by conducting a key informant survey in Besisahar, the administrative centre of Lamjung District. This enabled the research team to understand the broad context of remittance economy in the district, based on the district-wide information thus gathered. The research approach focused on understanding experiences and responses of local people and other actors in the state and functioning of CF and its nexus with labour out-migration for remittances.

Forty purposively selected informants were interviewed, including CFUG committee members, government officials, school teachers, political leaders, women leaders and the households directly involved in the establishment and operation of CFUGs. During the first round of the fieldwork in 2013, 23 key informant interviews were conducted, which included five at the district level and 18 at the local community level. The second field visit was in 2014, during which a household questionnaire survey was completed with 150 households. The selection of households was based on stratified random sampling, representing different socio-economic groups. During this stage, four workshops were held: one each with a)

CFUG committee members, b) local leaders including school teachers, c) women and d) Dalits. Of the total 40 interviews conducted, 25 interviews were with local communities, five were with district officials and NGO representatives, and 10 with government officials and NGO/INGO representatives at the national level. In 2016, two workshops were organized to share key findings of the research with and feedback obtained from local communities, government officials and NGO/INGO staff. The nature of the questions for different actors was focused on issues of CFUGs, remittances, labour out-migration and changes in the functioning of CFUGs, amongst others. The concurrent method of interviewing at district, VDC and community levels enhanced the research team's ability to refine questions and to fully contextualize the inquiry. Through the extended period of research, the research team explored the ways in which CF can manage and adapt to changing contexts and become more resilient.

Community forestry in the broader context of Nepal

Community forestry in Nepal is affected profoundly by volatile politics, fragile geography, expanding economy and unstable policy environment. The country has undergone massive political upheaval throughout its history, and more so in the last two decades. The politics of the country are unstable due to the frequent changes of the central government and deep impacts of the Maoist insurgency (1996–2006), coupled with the Royal Massacre in 2001 (Hutt 2004) and social paralysis due to ongoing constitutional crisis, the unfortunate rise of identity politics creating social divisions, and a stalemate in state restructuring (Paudel 2016; Satyal et al. 2017). The long insurgency and political reckoning has had a profound impact on the country's society, economy and environment (including effects on the functioning of CF). Key political events were the dissolution of the parliament by the king in 2005, and an extraordinary unfolding of events in Nepalese history including the abolition of the monarchy in 2007. Ongoing tensions within and between the major political parties led to the de-railing of the constitution-writing process for an extended period of time. The constitution was at last ratified by the parliament in September 2015, paving way for a new structure of federal states. Nepalese society and economy have been severely paralyzed by the continuing struggle with identity and ethnic politics centred on the issue of state restructuring. The new structure will consist of seven states, 75 districts with service roles but no governing role, and a new form of elected local government. The previously established administrative division of the country into five regions, 75 districts, 217 municipalities and 3,157 Village Development Committees (VDCs), on which many CF institutions are organized and functioning, has been changed to four metropolises, 13 sub-metropolises, 246 municipal councils and 481 village councils for official work (The Kathmandu Post, March 15, 2017). The exact number, boundaries and functional mechanism of the new states within the new federation are unclear, leading to significant confusion, which certainly affects CF. Such volatile political processes have severely hampered economic development. Nepal has become a vibrant democracy in political terms but it can also be seen as almost a failed or anarchical state. CFUGs are highly politicized, creating deep divisions and conflicts in the way CF is seen, institutions developed and forests managed.

While the constitutional debates and state restructuring continued to paralyze the country, Nepal's fragile geography was exposed, with the devastation from large earthquakes on 25 April 2015 (7.8 on the Richter scale) and 12 May 2015 (7.3 on the Richter scale). Over 500,000 houses were destroyed, 250,000 damaged, 8,790 people killed, 22,300 injured, and over 8 million (almost a third of the total population) affected (NPC 2015). The damage was estimated to be about $7 billion (NPC 2015). The ineffective government response was to implement prompt rescue operations, assess the damage in rural areas and coordinate national and outside support. International rescue operations were primarily concentrated in urban centres, and aid agencies and humanitarian organizations were incompetent or unable to execute comprehensive, coordinated and inclusive rescue and rehabilitation operations. Local communities and neighbourhoods, however, were the first to provide effective help in both rescuing the victims and helping each other in terms of immediate rehabilitation (Manandhar 2016). Many CF members were directly affected by the earthquakes and CF is seen as the source of timber specifically focusing on the rehabilitation and reconstruction of those mostly affected by the earthquake (Ojha et al. 2017).

The Nepal government has been receptive of the need, but the major concern has remained on meeting additional timber demand and CFUGs have sought technical advice on how they can obtain more timber than usual. Legal complexities and fear of legal action, however, remain an issue for the CFUGs. While the availability and use of timber from CF makes good sense during the time of disaster such as this, no decision has been made by the government to facilitate CFUGs to be able to sell timber in the market or to the needy households beyond the provisions made in operational plans, which make no consideration of such disasters. Many CFUGs are frustrated at not being able to mobilize timber available in the nearby forests to rebuild their houses. In this sense, the role and possibilities of CF during the time of disasters and recovery has become a matter of critical debate.

Nepal's forestry (and CF) is now linked with improving livelihoods and alleviating poverty. This makes sense as the country has significant success in protection of forests through CF and many community forests can be commercially utilized (Gilmour 2016a). Yet, the country is one of the poorest countries in the world with a per capita income of USD690 (World Bank 2015), ranking 109 of 195 countries on economic parameters of the World Bank with a low Human Development Index (HDI) of 0.548 (World Bank 2015). However, Nepal has managed to halve the percentage of people living on less than $1.25 a day in only seven years, from 53 percent in 2003–2004 to 25 percent in 2010–2011 (World Bank 2015). Recent World Bank data shows that Nepal's economic growth is 7.5 percent per annum, ranked third in a ladder of the fastest growing economies in the world, just below Ethiopia and Uzbekistan, and above India (World Bank 2017). The Nepal government aims to graduate from the least developed countries by 2022 (NPC 2013). The poverty is more severe in rural than in urban areas and more in the hills than in the Terai. Approximately 70 percent of the population live in rural areas, their livelihoods dependent on agriculture, forest and natural resources. Despite recent reports of declining poverty (from about 42 percent in 1995/96 to 25.4 percent in 2009 (NPC 2013), poverty levels remain high, as does socio-economic inequality (NPC 2013). This is evidenced by the fact that 3.48 million people suffered from severe food insecurity in 2011 (Dhakal, Bigsby and Cullen 2011). Agriculture contributes over 35 percent to GDP and forest lands about 10 percent (GoN 2013). While only 2.35 million hectares of the land is arable (World Bank 2005), more than 80 percent of the population depends on subsistence farming. Hence, agriculture is the mainstay of over 66 percent of the population and constitutes over 35 percent of Gross Domestic Product (GDP) (World Bank 2015). Sixty-nine percent of landowners in the country own less than one hectare of land (Bhattarai, Dhungana and Kafley 2007). With pressing challenges of poverty and need for sustainable use of forests, the question becomes how forests are managed under CF and elsewhere in such a way that dual purposes of poverty alleviation and sustainable forest management can simultaneously be achieved.

Nepal remains heavily dependent on development aid. About a third of the government's development budget is financed through foreign aid, which is further bolstered by funds separately channelled through non-governmental and direct aid channels. An estimated 70 percent of Nepal's national budget comes from foreign assistance (Tamang 2011). There is no comprehensive aid database in Nepal, and government data tends to underestimate the amount of money, as it captures only the formal and direct budgetary support (Devkota 2011). According to the Aid Cooperation Report (NPC 2013), foreign aid accounted for 6.2 percent of GDP in Fiscal Year 2012–2013 (with over USD1 billion in value), with 22 percent of the government budget coming from aid (53 percent and 47 percent respectively from loans and grants). In the context of the strong aid presence in Nepal's development, a specific program such as CF development in Nepal has received about USD237 million from donors over 30 years while the contribution from the Government of Nepal (GoN) over 20 years has been only around USD8 million (GoN 2013). Questions are being asked about the intent and procedure of designing and implementing donor-funded projects, such as the ten-year, over-$150-million-budgeted Multi Stakeholder Forestry Programme (MSFP) supported by the governments of Finland, Switzerland and the UK, which was dropped in 2015, seemingly due to the lack of willingness and commitment from Nepal's Ministry of Forest and Soil Conservation (Anon, pers. comm., 2016 with first author). Yet, Nepal's policies and programs including CF are largely donor driven and donor dependent.

With foreign aid support over a long period of time and also commitment from the government and local communities, Nepal's forestry sector has evolved with a progressive legal and policy framework developed over the past four decades. Nepal's forests are owned by the state and are managed under three broad management regimes: protected areas, government-managed forests and community forestry (CF). The Forest Act 1993 and the Forest Regulations 1995 provide a framework for regulating and managing all forest areas, under the broad guidance of the 1989 Master Plan for the Forestry Sector (MPFS) – a policy which expired in 2010. Nepal's forestry sector has undergone significant changes and the most notable change is the nationwide expansion of the CF program, which has become a key development success in Nepal, at least in terms of rehabilitating the once degraded forests in the hills (Niraula et al. 2013; Pandey and Paudyall 2015; Pokharel, Tiwari and Thwaites, Chapter 3 in this volume). At the same time, Nepal's forest sector has also faced new challenges, most of which relate to balancing multiple and conflicting forest values (Paudel 2012) – such as biodiversity, livelihoods, watershed protection and, increasingly, carbon sequestration – in changing global and national contexts. Following the expiration of the MPFS (1989–2010), Nepal's Ministry of Forest and Soil Conservation (MFSC) has drafted a new Forestry Sector Strategy for the next decade (2015–2025) (GoN 2013). The policy is awaiting approval from the Government of Nepal. Once approved, the strategy is to serve as the successor of MPFS to guide the forestry sector. However, the policy has allegedly been controversial within the forest bureaucracy itself. It has been criticized for ignoring scientific evidence on forest policy and practices available in Nepal and internationally, and also driven by the interests of a small network of donors and some MFSC officials without broad government participation and ownership (Ojha et al. 2015). The recent discourses of 'forest for prosperity', 'sustainable forest management' and forestry in the age of climate change and REDD+ are all in vogue with Nepal's senior forestry officials and are being forcefully incorporated in the forthcoming 14th National Plan with little deliberation by CFUG membership around the country (Anon, pers. comm. with first author 2017). The fate and future of CF as a locally controlled forest management regime remain highly uncertain.

While political and natural disasters and their impacts and recovery efforts continue, Nepal is experiencing unprecedented out-migration in recent years and remittances have become one of the major sources of income for Nepal. As a result, males are moving out of the agriculture field, leaving females to manage the farms. This trend is ever increasing (Ojha et al. 2017; Satyal et al. 2017). There is an increasing trend of people migrating temporarily and permanently outside of their home country in search of a better life (Paudel, Tamang and Shrestha 2014). According to the national population and housing census of 2011 (CBS 2012), over 24 percent of all households reported at least one immediate family member was living away from the household, and the number of absent people has grown rapidly from 3.3 percent of the total population in 2001 to 7.3 percent in 2011. The Central Bureau of Statistics reports that, in 2011, 56 percent of households in Nepal were receiving remittances (CBS 2011), while ADB figures indicate that 31.5 percent of the national GDP was derived from remittances in 2015 (ADB 2016).

With globalization and advances in communication and transportation, many people, particularly adult men in developing countries, are migrating temporarily and permanently outside of their home country in search of a better life (Bhadra

2007). The important fallout of this migration is that agricultural labour is being increasingly feminized (Cornhiel 2006; Gartaula, Niehof and Visser 2010; Kollmair and Hoermann 2011). This is happening with the growing awareness of and an interest in formal education in all families; children are going to school and young people in particular are no longer seen in the agriculture fields, having been lured by foreign employment. The burden of both farming and looking after the household has fallen on the shoulders of women, in what can be described as the 'feminization of agriculture'. In the absence of male counterparts, the female members of the households are taking more responsibility for carrying out agricultural activities within the traditionally male dominated agricultural system. This is not only adding extra workloads to women but it is also culturally inappropriate and often uncomfortable for them.

The changing context of CF: the rise of the remittance economy

Community forestry in Nepal has undergone unprecedented change in recent years due to the emergence of the remittance economy and emigration of labour overseas. All households surveyed in the field research noted that out-migration has affected the functioning of CF. This was amplified by the majority of interview respondents and workshop participants. A senior forester summarized some major issues as follows:

> With the advent of opportunities overseas, limited opportunity at home and ongoing political infighting in Nepal, many youth prefer to go overseas as soon as they can. They often ignore the risks of underpayment and difficulties at work. Our airport is full of people traveling to and from the Gulf countries every day, while only women, children and elderly are left in the villages. . . . The active manpower has been exported elsewhere, while the development of the country is in a dire state. This is the biggest paradox facing Nepal's development.
> . . . And much lauded community forestry institutions at the local level are struggling to function because many people are moving out of the villages. No one seems to be interested to run the committee. . . . People value forests less and less because they have income from overseas. They can now buy foods and timber products from the market. With the highly politicized local environment in Nepal, community forestry is now truly at a crossroad.

The statements above resonate with many people's views. Despite such unprecedented change at the local and national levels, academic and policy debates on community forestry do not include the effects of the remittance economy in Nepal's villages. As such, CF policies and practices fail to engage with, and respond to, the broader social, political and economic changes brought about by the remittance economy. The remittance economy has no doubt become massive and it directly affects the way CF has functioned. The question then is how and to what extent CF practice has been affected by it. The research shows that the remittance economy has become a major threat to the continuing functioning and sustainability of CF for four main reasons; a) out-migration, feminization and institutional instability; b) decreasing forest dependency; c) shifting social values; and d) lack of strategic and inclusive CF planning.

Out-migration, feminization and institutional instability

The remittance economy has profound social ramifications. With many men having moved away from the villages, local communities are feminized. In many cases, members and committee members of CFUGs are all women. Sixty-three percent of household respondents reported that CF institutions are now less active because women members have less time for CF activities because of their multiple and often-competing responsibilities within and outside the household. Further, 85 percent of household respondents stated that women-only groups are struggling to run CFUGs because they do not have sufficient experience or capacity. Despite the opportunity for women to make communal decisions in CF, the majority of survey respondents noted that women are not enthusiastic to attend CF assemblies or run Executive Committees. It was revealed that mostly women or elderly men attended community workshops.

Most workshop participants said that local communities are facing challenges to protect forests because of the increased access to markets. Many new roads are coming through the villages and illegal cutting of timber by outsiders has been frequently reported. When asked about the increased use of community forests, most workshop respondents said they were not confident that the CFUGs could manage the increased use. These views from the local communities are contradicted by the views of national-level actors. For instance, workshop participants at the national level highlighted the need for active utilization of community forests. This mismatch of views indicates the gulf in expectations amongst local- and national-level actors concerning forest use.

A major point coming out of the research was that there is a high level of emigration due to overseas employment and subsequent movements of people, and this has created an institutional and leadership vacuum within CF groups. As a result, CF is less stable and dynamic, as one of the central leaders of FECOFUN (Federation of Community Forestry Users, Nepal) said:

> Local communities, primarily women household heads, are leaving community forestry not because they do not love forests, but because they do not have sufficient time, interest, manpower or capacity to function as members of active forest user groups.

A school teacher highlighted institutional vulnerability in the face of political challenges and said:

> The forest users committee is vulnerable to political party interest because most people who had a long interest in community forestry have either

retired, or are not interested, or have left the areas. Women are less interested or capable to deal with the politically charged positions.

The issue of institutional fragility was also raised by a government officer at the
district level who said

> Community forestry groups are no longer as active as they used to be. Many
> operational plans are not renewed. There is in fact no demand for renewal.
> People tend to practice passive forest management. Out-migration has cre-
> ated this mess.

Decreasing dependence of people on forests

With the emergence of the remittance economy in Nepal's villages, one visible
phenomenon is that many local people do not often go to the forests to collect
fodder, fuelwood and timber. Over the period of four years, researchers were con-
sciously looking for evidence as to how often people go to the forests and why.
Based on informal conversations, it was clear that people do not need to go to
the forests that often. They cook food with the natural gas available at the local
market, which replaces the need for firewood to cook meals. Livestock raising has
significantly decreased (thus reducing the need for fodder). Markets have developed nearby and people can buy wooden and aluminium furniture. Seventy-nine
percent of survey respondents agreed that people's dependence on community
forests has dramatically decreased. When asked if such a decrease in dependence
was mainly due to remittances, the majority of people agreed. It was a common
view that forests are no longer an integral part of livelihoods. As one local community member said:

> Community forests are less important. We do not need firewood except dur-
> ing the festivals. We do not need fodder. We do not keep many buffaloes,
> cows and goats. Meat, milk, and eggs can be bought from the local market.
> If you have money, you can buy almost anything. Most households have at
> least one person in Arab [countries], so they have money to buy these prod-
> ucts. Raising livestock is difficult. If you can buy the products you want, why
> do you bother to do the hard work?

An NGO worker stated, "many people have left raising livestock and cultivating
land in recent years because women, children and the elderly simply cannot do
these jobs. Villagers have alternatives. They can buy the same products in the
local market."

When asked about the role of community forests in the local economy, workshop participants at the community level said that CF does not contribute much
because community forests are protected, not utilized. As mentioned above, local
communities are not necessarily interested in or confident about increasing the
use of forests. However, the need for active utilization was raised as a critical issue

at the national-level workshops, and by actors at the national and district levels. One senior government officer said:

> Passive forest utilization is the main reason why local people are less interested in forest management. Many community forests are matured but they are not utilized because local communities have a very restricted operational plan. There is a need for sustainable forest management.

Many forest officers and NGO workers have highlighted the problem of subsistence-based forestry and forest under-utilization, whereas local people seem to be less keen on active forest utilization. However, at the national level, the climate change and biodiversity conservation discourse, focused mainly on biophysical science, has reinforced the previous policy discourse favouring forest protection. The protection-focused discourse has changed, at least amongst forestry policy decision-makers, to an emphasis on sustainable forest management, which highlights the need for active and sustainable use of forests. Thus, dual contradictions exist: one is the contradiction between decreasing interest in active forest management at the local level and active forest management discourse at the national level amongst forest policy decision-makers; the other is the contradiction between the sustainable use discourse of the forest policymakers and the conservation focus of the climate change and biodiversity discourse. These contradictions have created a stalemate in community forest management practice and reinforced the status quo.

A question was asked about the relationship between the dependence on forests and number of people going overseas for jobs. All household survey respondents noted that people are less dependent on forests because they have money to buy alternative or similar products from the markets, which they would have otherwise obtained from community forests. It was clear that the nature of the community forest relationship is rapidly changing. Many local people have different perceptions on the value of forests – the forests are seen less as a direct source of livelihoods, more as natural resources 'nice' to be conserved. This is a significant change, and a change that needs to be recognized by CF policies and legislation so as to maintain CF as a relevant and useful forest management approach in the future.

Shifting values of society and the priority on wealth maximization

The traditional Nepali society based on caste is changing rapidly. A new social priority has emerged on wealth maximization, which is brought about not only by the remittance economy but also by the expanding capitalistic political economy, the abolition of the monarchy, entry of the radical communists into the political mainstream and the effects of globalization. The research indicates that Nepali households and communities are more open, traditionally marginalized groups more empowered, educational level of households increased, aspirations of youth more ambitious and the markets rapidly developed. Amongst these changes, respondents highlighted that wealth maximization has become the number one priority for Nepali households and individuals, whether a politician, a Dalit or a teacher. One teacher said:

> Wealth has taken over the caste system to determine social prestige. If a Dalit family has wealth, that family is well regarded in society. On the other hand, if a Brahmin family is poor, that family will have low social prestige. This is the new normal.

Respondents answered that the main source of wealth in Nepali communities has become remittances. In fact, remittances are the major source of revenue for the national economy. One political leader discussed the fate and future of the Nepali economy in the following terms:

> Nepal is dangerously dependent on remittances. It's more than the foreign aid. Many poorer families are coming out of poverty because of remittances. This is good, but it is a risky source of income. Who knows that with some change in the policies of another country, the whole household and national economies of Nepal can come down to their knees. The national discourse of the remittance economy being good therefore needs careful consideration. . . . Most remittances have been spent on purchasing houses, land and on household consumption. It's extraordinary that people do not invest remittances to set up and run businesses. Reasons are many but mainly due to not having conducive political and fiscal environments. . . . Remittances are not spent on community forestry. In fact, the rise of the remittance economy has threatened the participation of local communities in forest protection.

Seventy percent of household survey respondents noted that they would need to earn more money to maintain the changing life style based on the market economy. They stated that traditional farming is no longer viable because a whole year of farming does not even feed the family for half a year due to the low productivity of land. Farming is hard, and seen as a low-prestige profession in society. Hence, land under-utilization is rampant. Ironically food security is growing while the available land is not used. When asked why people are leaving the land fallow, one elderly women leader said:

> Farming is hard and expensive. Bulls, labour and fertiliser are very expensive. Further, we do not find many people who want to work on farms. Most of us only cultivate a small plot of land around our houses for vegetables. As the roads are coming everywhere, many people are dividing land for household parcels and selling it to new migrants. We cannot find good tenants. Everyone is waiting to fly out to the Arab [countries]. Young people do not love the land that much.

Many interview respondents said that while people are abandoning private land, communal land under community forestry is not a big concern. People do not often talk about what to do with community forests because individual opportunities and challenges are taking over communal matters. The clash between collective and individual values, and the priority for wealth maximization, has seriously affected the present and future of community forestry.

Exclusionary CF institutions in the midst of prolonged political transition

The research indicates that local CF leadership is exclusionary. Eighty-seven percent of household survey respondents agreed that Executive Committee members are village leaders, and the ones who make CF decisions. One focus group involving only Dalits highlighted that they are interested in and dependent on forests, but the existing institutions and practices have restricted them from being more active in the management of CF. One NGO officer summarized the current state of exclusionary planning within CFUGs:

> Local community forestry rules continue to be captured by elites – elderly male or elite female. These elites are less interested because they are less dependent on forests. Often, the quorum for the meeting is not fulfilled. Still they occupy critical positions and maintain the status quo. On the other hand, many Dalit groups are more dependent on forests, but they are not given meaningful responsibility for protecting the forest. Forest managers are those who are less dependent and less interested, while those having critical stake on forests are marginalized by the existing CF institutions.

One academic who has engaged with many community groups in Nepal over the last 30 years highlighted the need to consider an alternative planning system for community forestry at the local level:

> The operational planning process for community forestry is over 40 years old. This system is still being repeated across the country with some success. However, politics has changed, the economy has changed, society is very different now. This planning system needs to be reviewed. Further, this planning is about forest protection. Annual planning is what is being practiced. There is no strategic planning of how to manage the forests in the next ten or twenty years time. It is long overdue – the community forestry planning system needs to be reconsidered in view of changing society and environment.

Local- and national-level workshops revealed that there is a limited demand for local democracy, as no local government election had been held in 16 years. Respondents highlighted 'bigger' issues such as state restructuring, constitutional crisis, geopolitics and political controversies – almost everyone thought community forestry does not demand attention in the current political climate. When asked about the fate and future of CF in the changing political environment, one of the prominent political leaders of a major party summarized his views of the situation:

> The country is going through a difficult time. A constitutional crisis has unsettled the country for a long period of time. Identity politics has divided the society; geopolitics involving India's and China's interest in Nepal has never been so rampant. Political parties seem to disagree about almost everything. Over 1500 youth are leaving the country every day for work in the Gulf. Development aid agencies have been unaccountable. The debate on state restructuring is going nowhere as political leaders do not agree on numbers and boundaries of the state. The leadership is paralysed and incompetent, creating significant uncertainty for the life and livelihoods of everyday citizens. In this massive pool of problems, unfortunately, local innovations such as community forestry are on the shadow.

Revitalizing community forestry in the changing national and global context

The findings presented above show that smooth functioning and sustainability of CF is threatened with the rise of the remittance economy in Nepali villages. Youth out-migration has become a key causal force to threaten the sustainability of CF. This is linked not only to the need for cash incomes, but also to aspirational factors in which moving out of an 'underdeveloped' region is seen as a symbol of social status. The rise of the remittance economy is widespread. This change has brought new livelihood strategies, which are Nepal-wide. The case studies reveal that migration contributes to CF management in two ways:

- first, due to its direct effect on the supply of manpower for forest management; and
- second, due to the indirect effect of remittances in providing villagers with the option of moving out of community forestry.

Then, a critical question for CF scholars and practitioners is how to understand multiple forces that potentially jeopardize the functioning and sustainability of CF, and how to revitalize CF policies and practices in such a way that CF institutions adapt effectively to the new environment and make the best use of opportunities emerging in response to local, national and international contexts.

In this section, we explore the important implications for the fate and future of CF in Nepal. We also advance the discussion towards reinforcing the need for an inclusive and strategic approach to planning of CF and highlight the need for institutional reframing of CF within the changing context. The evidence presented here shows that it is not always sensible to focus only on the causes and impacts of the remittance economy on CF, which include out-migration, feminization and institutional instability, decreasing forest dependency, shifting social values and lack of strategic and inclusive planning, as discussed in the preceding sections; rather, attention should also be focused on how such impacts can be managed. It is therefore useful to go back to our key questions: how has the remittance economy changed the environment within which CF operates, and whether and how the remittance economy helps (or hinders) the progress of CF? Clearly, the remittance economy has become a threat to the progress of CF for the various reasons explained above. The question is how Nepal's CF can remain resilient in the changing national and global contexts? Below we discuss some ways as to how Nepal's CF can manage the threat coming from the remittance economy.

Enhancing strategic and inclusive planning practice in CF

The analysis of Nepal's remittance economy in relation to CF policies and practice reveals that there has been a very limited opportunity for local communities, particularly the Dalits and poorest of the poor, to effectively participate in the CF planning process. In fact, the CF planning process was designed some 40 years ago and the same process continues today. Moreover, the planning is limited to annual operational plan making – a process which is overly dependent on government foresters. No longer term strategic planning has been part of CF. This explains why CF fails to anticipate and manage changes such as those resulting from the remittance economy. This is not to say that there has been no planning or opportunity for participation at all. Rather, the opportunities provided are limited to a few individuals, use universal methods of engagement, implemented in haste, and have ignored the diversity of communities. The existing planning process and the opportunities for participation are, therefore, not sufficient and suitable to bring about effective broad-based participation of communities as contributors and users of knowledge about planning and policy issues as envisaged by Lewenstein (2004). The underlying issue here is essentially about the custodial culture and scientific tradition of the top-down, narrowly conceived and out-dated planning processes in CF (Shrestha 2012). Clearly, there is a need for strategic planning processes that facilitate deliberation and inclusion of otherwise sub-ordinated voices, and the plans that can anticipate and manage changes in society, economy and environment. In other words there is need for a more inclusive planning process, encompassing affirmative action that provides opportunities for the voiceless to participate in planning and decision-making in ways that affect their interests now and in the future.

Yet, the wider project of modernization in Nepal has consistently failed in terms of strategic and inclusive planning. This resonates with Rigg's *Unplanned development*, in which he highlights the "gaps between planning designs and planning experiences, between what is seen and measured and what ultimately proves to be important, and between expectations and outcomes" (Rigg 2012, p. 3). Nepal recently (2016) launched its 14th Periodic Plan, aiming to foster modern development, but in practice there exists a yawning disconnect between planning designs and actual development practice. The state, in its planning pronouncements, has always proclaimed the value of forestry and farming, but has hardly considered the effects of the remittance economy involving the dynamics of migration, economic inter-linkages, infrastructure development and processes of modernization, all of which have contributed to the weakening of community forestry institutions.

Adaptive and collaborative learning

A growing body of literature has delved into how policy and institutions emerge, function, change and improve to address the challenges such as those from the remittance economy (Colfer 2005; Shrestha 2016; Ojha et al. 2017). More operationalizing concepts of learning and innovation have emerged from social and organizational learning fields (such as Schon 2010), and on works that emphasize integrated analysis of society and natural systems (Holling 2001). We identify at least three overlapping factors through which CF could become more resilient in the face of the remittance economy. First, dominant CF actors could become more reflective in view of challenges and opportunities brought about by changing remittance economy contexts. Second, locally engaged critical intellectuals and researchers can help in the generation of alternative evidence, alternative facts and knowledge, as found in the case of slum dweller empowerment work in the Indian city of Mumbai (Patel, Baptist and D'Cruz 2012). Similar work on Nepal's forest sector also demonstrates the potential of critical action research in exposing conventional policy assumptions and critiquing the conventional practice (Ojha et al. 2015). The practice of critical action research in forestry (e.g. Ojha 2013) has the potential to simultaneously enable local communities and forest authorities to recognize the value of local knowledge and limits of dominant knowledge systems and to generate critical evidence to stimulate new thinking to anticipate and manage wider changes. Third, employing an adaptive approach to learning and knowledge production can support the management of forests in the face of uncertainty and multiple challenges. Adaptive and collaborative approaches provide the space to relevant actors, at different levels, to interact and negotiate the goals of resource management and to innovate institutional mechanisms that provide better ways of managing resources to meet the changing context (Ojha et al. 2013). An adaptive collaborative management approach has been tested by CIFOR (McDougall et al. 2007).

Enhancing the knowledge interface practice

Multiple forms of knowledge are useful to improve policies and practices of community forestry. Local people, policymakers and scholars have valuable knowledge and experience of how to improve forest use and management, but the existing platform of CFUGs does not offer an effective platform for a diversity of local communities, particularly Dalits and poor groups, to share their knowledge with government officials and scholars, and to influence local practices and policies in the changing context. Local elites continue to make use of their knowledge for local decisions, while forestry officials develop policies based on what they consider to be important forestry problems and solutions. Knowledge of the 'poorest of the poor' and Dalits is effectively excluded from the CFUG process. While there is a perception that the CFUG is a good platform for engaging local communities, it has not been proven to be an effective platform for sharing the knowledge of the poor and disadvantaged, engaging them in decision-making and empowering them (Shrestha 2016).

The question, then, is what could potentially work for different groups of stakeholders within the local community, and between the local community and government officials to share their knowledge more effectively and contribute in making CF more resilient in a changing context. This is essentially about the knowledge interface, which can be defined as a process by which different groups of people meet and share their knowledge, as well as educate each other. Since the existing CFUG is an established institution, but it is not facilitating effective knowledge interface, it is useful to consider how to enhance the knowledge interface in the existing institutional platform to make the best use of knowledge held by different stakeholders. An enhanced knowledge interface has the potential to help government agencies, NGOs and local communities to make informed decisions in improving institutions and developing inclusive plans (e.g. Ojha 2013; Shrestha 2016).

The question of how exactly the enhanced knowledge interface is likely to work in practice is beyond the scope of this chapter, but it is useful to highlight why an enhanced knowledge interface is important in revitalizing CF. First, a knowledge interface can catalyze different knowledge holders such as Dalits, poor users, women, business people, scholars and forestry officials to be able to share, listen, understand, negotiate and change their understandings of CF in the context of broader changes such as the rise of the remittance economy. Instead of assuming how different groups are affected or would experience the change in economy and CF, the interface would directly bring about the voice and experience of people at the forefront, which could lead to developing practical ways to offer targeted support to CFUGs and forestry officials. It has been argued that through the synthesis of different forms of knowledge by engaging different knowledge holders – from local communities to scientists and policymakers – an enhanced knowledge interface could help strengthen community risk reduction responses (Weichselgartner and Kasperson 2010; Gaillard and Mercer 2013). Second, a knowledge interface would facilitate local communities such as CFUGs to share grounded insights, while good practices could be taken up by the policy actors such as forestry officials (e.g. Shaw, Uy and Baumwoll 2008). Finally, a knowledge interface can be a vehicle for rethinking the role of experts, policymakers and local community in environmental governance (see Fischer 2000; Backstrand 2003), which could benefit the revitalization of CF by raising the profile of previously sub-ordinated voices and experiences of some local groups.

While specifying explicit solutions is beyond the scope of the chapter, we believe that the pathways approach to understanding remittance economy highlights a few important aspects that could be considered in CF planning and institutional transformations. First, the nature of the remittance economy pathways – involving multi-scalar (involving local, national and global forces) and cross-sectoral (covering social, environmental, political and cultural) processes – means that no single policy intervention can make CF resilient in the face of unprecedented changes in society and economy. Second, in the wider policy question of agrarian development, resource governance and sustainable development goals, it is even debatable whether the solution should involve keeping the present form of CF. The specific struggles of CF institutions are only a symptom of larger policy failures, and a policy inquiry may consider such patches as only tiny problems in the larger landscape of resource governance, migration and development. Any policy solutions must recognize the changing expectations of society in Nepal, while elevating forestry, livelihoods and development as the important national agenda. It is not just the functioning and sustainability of CF that is at stake here, but the fundamentally shifting relationships amongst people, culture, land and livelihoods in the rapidly globalising world. The policy response itself should be grounded in the varied pathways that lead to different contexts and the extent of threats to CF in different localities. Changing context is a critical consideration. More critical pragmatic research is needed to explore context-specific socio-institutional options that match ecology, local economy and culture, as well as a process of strategic and inclusive planning that draws on research, local knowledge and views of the people who are most disadvantaged in the process.

Conclusion

In this chapter, we have analyzed the fate and future of CF in the context of the remittance economy in Nepal. The chapter has highlighted unprecedented changes brought about by this remittance economy to Nepal's rural villages and towns. Local institutions such as Community Forest User Groups and the management of the environment through programs such as CF have been profoundly affected. We have discussed how and to what extent the effects of the remittance economy are felt in CF, and what future lies ahead for CFUGs and the sustainable management of CF. We argue that the remittance economy is a threat to the sustainability of CF. For CF to continue advancing in the new context of neoliberal political economy, it needs to be reframed in such a way that its planning process is modified and updated, and CF institutions need to consider the remittance economy as an opportunity to actively utilize forests for equitable development and to see markets as a platform to help alleviate poverty. We argue that CF needs to adapt to the new economic environment through improving planning processes and institutional reframing. In doing so, we have also identified how CF planning and institutional reframing can be advanced by politicizing local and scientific knowledge. Clearly, there is a disconnect between the amplified narrative of remittance as a national 'good' without critically understanding and addressing social and environmental issues at the local level. This requires major rethinking in the way the remittance economy is being used in global and national CF policy and practice reforms.

Note

1 The project was supported by an Australian government's Endeavour Research Fellowship.

References

ADB 2016, *Asian Economic Integration Report 2016: What Drives Foreign Direct Investment in Asia and the Pacific?* Asian Development Bank, Mandaluyong City, Philippines.

Agrawal, A., Chhatre, A. and Hardin, R. 2008, Changing governance of the world's forests, *Science*, vol. 320, pp. 1460–1462.

Backstrand, B. 2003, Civic science for sustainability: Reframing the role of experts, policy-makers and citizens in environmental governance, *Global Environmental Politics*, vol. 3, no. 4, pp. 24–41.

Bhadra, C. 2007, *International labour migration of Nepalese women: Impact of their remittances on poverty reduction*, Working Paper Series No. 44, Asia-Pacific Research and Training Network on Trade, (ARTNet), Bangkok.

Bhattarai, B., Dhungana, S.P. and Kafley, G.P. 2007, Poor-focused common forest management: Lessons from leasehold forestry in Nepal, *Journal of Forest and Livelihood*, vol. 6, no. 2, pp. 20–29.

CBS 2011, *Nepal Living Standards Survey 2010/11, Statistical Report Vol. 2*, Central Bureau of Statistics, Government of Nepal, Kathmandu, Nepal.

CBS 2012, *National Population and Housing Census 2011*, Central Bureau of Statistics, Government of Nepal, Kathmandu, Nepal.

Cedamon, E., C., Nuberg, I., Pandit, B. and Shrestha, K.K. 2017, *Adaptation factors and futures of agroforestry systems in Mid-hills of Nepal*, Agroforestry Systems, doi:10.1007/s10457-017-0090-9.

Colfer, C.J. (ed.) 2005, *The Equitable Forest: Diversity, Community and Resource Management*, Resources for the Future, Washington, DC.

Cornhiel, S. 2006, *Feminization of agriculture: Trends and driving forces*, Background Paper for the World Development Report, Rimisp-Latin American Center for Rural Development.

Devkota, J. 2014, Impact of migrants' remittances on poverty and inequality in Nepal, *International Development Research Forum*, vol. 44, pp. 36–53.

Devkota, P. 2011, *Changing Paradigms of Aid Effectiveness in Nepal*, Alliance for Aid Monitor Nepal, Lalitpur.

Dhakal, B., Bigsby, H. and Cullen, R. 2011, Forests for food security and livelihood sustainability: Policy problems and opportunities for small farmers in Nepal, *Journal of Sustainable Agriculture*, vol. 35, no. 1, pp. 86–115.

DoF 2017, *Community Forestry Status Database*, Department of Forests, Kathmandu, Nepal, viewed 26 May 2017, http://dof.gov.np/dof_community_forest_division/community_forestry_dof

FAO 1978, *Forestry for local community development*, FAO Forestry Paper No. 7, Food and Agriculture Organization of the United Nations, Rome, Italy.

Fischer, F. 2000, *Citizens, Experts and the Environment: The Politics of Local Knowledge*, Duke University Press, Durham, NC and London.

Gaillard, J. and Mercer, J. 2013, From knowledge to action: Bridging gaps in disaster risk reduction, *Progress in Human Geography*, vol. 37, no. 1, pp. 93–114.

Gartaula, H.N., Niehof, A. and Visser, L. 2010, Feminisation of agriculture as an effect of male out-migration: Unexpected outcomes from Jhapa District, Eastern Nepal, *The International Journal of Interdisciplinary Social Sciences*, vol. 5, no. 2, pp. 565–578.

Gilmour, D. 2016a, Unlocking the wealth of forests for community development: Commercializing products from community forests, in J. Meadows, S. Harrison and J. Herbohn (eds), *Small-Scale and Community Forestry and the Changing Nature of Forest Landscapes*, Proceedings from the IUFRO Research Group 3.08 Small-Scale Forestry Conference held on the Sunshine Coast, Queensland, Australia, 11–15 October 2015, pp. 78–93.

Gilmour, D. 2016b, *Forty years of community-based forestry: A review of its extent and effectiveness*, FAO Forestry Paper No. 176, Food & Agricultural Organization of the United Nations (FAO), Rome.

GoN 2013, *Persistence and Change: Review of 30 Years of Community Forestry in Nepal*, Ministry of Forests and Soil Conservation, Government of Nepal, Kathmandu, Nepal.

Holling, C.S. 2001, Understanding the complexity of economic, ecological, and social systems, *Ecosystems*, vol. 4, pp. 390–405.

Hutt, M. (ed.) 2004, *Himalayan 'People's War': Nepal's Maoist Rebellion*, Hurst, London, UK.

The Kathmandu Post 2017, 744 new local units come into effect, viewed 13 March 2017, http://kathmandupost.ekantipur.com/news/2017-03-15/744-new-local-units-come-into-effect.html

Khatri, D., Shrestha, K.K., Ojha, H., Paudel, N. and Paudel, G. 2016, Reframing community forestry governance for food security in Nepal, *Environmental Conservation*, vol. 44, no. 2, pp. 174–182.

Kollmair, M. and Hoermann, B. 2011, Labour migration in the Himalayas: Opportunities and challenges for mountain livelihoods, *Sustainable Mountain Development*, vol. 59, pp. 3–8, ICIMOD, Kathmandu, Nepal.

Lewenstein, B. 2004, *What does citizen science accomplish?* CNRS Colloquium, 8 June 2004, Paris, viewed 20 June 2017, https://ecommons.cornell.edu/handle/1813/37362

Manandhar, B. 2016, Remittance and earthquake preparedness, *International Journal of Disaster Risk Reduction*, vol. 15, pp. 52–60.

McDougall, C., Ojha, H., Pandey, R.K. and Pandit, B.H. 2007, Enhancing adaptiveness and collaboration in community forestry in Nepal, in R. Fisher, R. Prabhu and C. McDougall (eds), *Adaptive Collaborative Management of Community Forests in Asia: Experiences from Nepal, Indonesia and the Philippines*, Center for International Forestry Research (CIFOR), Bogor, Indonesia, pp. 52–92.

Niraula, R.R., Gilani, H., Pokharel, B.K. and Qamer, F.M. 2013, Measuring impacts of community forestry program through repeat photography and satellite remote sensing in the Dolakha district of Nepal, *Journal of Environmental Management*, vol. 126, pp. 20–29.

NPC 2013, *Thirteen Plan – Three Year Plan (2013/14–2015/16) Approach Paper*, National Planning Commission, Government of Nepal, Kathmandu, Nepal.

NPC 2015, *Post Disaster Needs Assessment*, National Planning Commission, Government of Nepal, Kathmandu, Nepal.

Ojha, H.R. 2013, Counteracting hegemonic powers in the policy process: Critical action research on Nepal's forest governance, *Critical Policy Studies*, vol. 7, pp. 242–262.

Ojha, H.R., Khatri, D., Shrestha, K.K., Bhattarai, B., Baral, J., Basnett, B., Goutam, K., Sunam, R., Banjade, M., Jana, S., Bushley, B., Dhungana, S. and Paudel, D. 2015, Can evidence and voice influence policy? Critical review of Nepalese forestry sector policy, *Society and Natural Resources*, vol. 29, no. 3, pp. 357–373.

Ojha, H.R., Khatri, D., Shrestha, K.K., Bushley, B. and Sharma, N. 2013, Carbon, community and governance: Is Nepal getting ready for REDD+? *Forests Trees and Livelihoods*, vol. 22, no. 4, pp. 216–229.

Ojha, H.R., Shrestha, K.K., Subedi, Y., Shah, R., Nuberg, I., Heyjoo, B., Cedamon, E., Rigg, J., Tamang, S., Paudel, K., Malla, Y. and McManus, P. 2017, Agricultural land underutilisation in the hills of Nepal: Investigating socio-environmental pathways of change, *Journal of Rural Studies*, vol. 53, pp. 156–172.

Ojha, H.R., Timsina, N., Kumar, C., Banjade, M. and Belcher, B. 2007, *Communities, Forests and Governance: Policy and Institutional Innovations From Nepal*, Adroit Publisher, New Delhi, India.

Pandey, G. and Paudyall, B. 2015, *Protecting forests, improving livelihoods – community forestry in Nepal*, FERN – Making the EU Work for People & Forests, viewed 20 June 2017, www.fern.org/sites/fern.org/files/fern_community_forestry_nepal.pdf

Patel, S., Baptist, C. and D'Cruz, C. 2012, Knowledge is power: Informal communities assert their right to the city through SDI and community-led enumerations, *Environment and Urbanization*, vol. 24, no. 1, pp. 13–26.

Paudel, D. 2012, In search of alternatives: Pro-poor entrepreneurship in community forestry, *Journal of Development Studies*, vol. 48, no. 11, pp. 1649–1664.

Paudel, D. 2016, Re-inventing the commons: Community forestry as accumulation without dispossession in Nepal, *Journal of Peasant Studies*, vol. 43, no. 5, pp. 989–1009.

Paudel, K., Tamang, S. and Shrestha, K.K. 2014, Transforming land and livelihoods: Analysis of agricultural land abandonment in the mid-hills of Nepal, *Journal of Forest and Livelihood*, vol. 12, no. 1, pp. 11–19.

Pokharel, B. 2011, *Green Governance: Development Aid, Local Investment and Benefits From Community Forestry*, Aid Monitor, Kathmandu, Nepal.

Rigg, J. 2012, *Unplanned Development: Tracking Change in Asia*, Zed, London, UK.

Satyal, P., Shrestha, K.K., Ojha, H., Vira, B. and Adhikari, J. 2017, A new Himalayan crisis? Exploring transformative resilience pathways, *Environmental Development*, vol. 23, pp. 47–56.

Schon, D. 2010, Government as learning system, in C. Blackmore (ed.), *Social Learning Systems and Communities of Practice*, The Open University Press, Milton Keynes, UK, pp. 5–16.

Shaw, R., Uy, N. and Baumwoll, J. (eds) 2008, *Indigenous Knowledge for Disaster Risk Reduction: Good Practices and Lessons Learned From Experiences in the Asia-Pacific Region*, United Nations International Strategy for Disaster Reduction, Kyoto University and the European Union, viewed 18 September 2016, www.unisdr.org/files/3646_IndigenousKnowledgeDRR.pdf

Shrestha, K.K. 2012, Towards environmental equity in Nepalese community forestry, in F.D. Gordon and G.K. Freeland (eds), *International Environmental Justice: Competing Claims and Perspectives*, ILM Publications, Hertfordshire, UK, pp. 97–111.

Shrestha, K.K. 2016, *Dilemmas of Justice: Collective Action and Equity in Nepal's Community Forestry*, Adroit Publishers, New Delhi, India.

Shrestha, K.K. and Ojha, H. 2017, Theoretical advances in community-based natural resource management: Ostrom and beyond, in G. Shivakoti, U. Pradhan and H. Helmi (eds), *Redefining Diversity and Dynamics of Natural Resources Management in Asia*, Vol. 1, Elsevier, UK, pp. 13–40.

Sunam, R., Paudel, N. and Paudel, G. 2015, Community forestry and the threat of recentralization in Nepal: Contesting the bureaucratic hegemony in policy process, *Society and Natural Resources*, vol. 26, no. 12, pp. 1407–1421.

Tamang, S. (ed.) 2011, *Looking at Development and Donors: Essays From Nepal*, Martin Chautari, Kathmandu, Nepal.

Tamang, S., Paudel, K. and Shrestha, K.K. 2014, Feminization of agriculture and its implications for food security in Nepal, *Journal of Forest and Livelihood*, vol. 12, no. 1, pp. 20–32.

Tiwari, S. and Bhattarai, K. 2011, *Migration, remittances and forests: Disentangling the impact of population and economic growth on forests*, Policy Research Working Paper No. WPS 5907, World Bank, Washington, DC.

Weichselgartner, J. and Kasperson, R. 2010, Barriers in the science-policy-practice interface: Toward a knowledge-action system in global environmental change research, *Global Environmental Change*, vol. 20, no. 2, pp. 266–277.

World Bank 2005, *World Development Report 2006: Equity and Development*, World Bank, Washington, DC.

World Bank 2015, *World Development Report 2015: Mind, Society, and Behaviour*, World Bank, Washington, DC.

World Bank 2016, *Migration and Remittances Factbook 2016*, 3rd edn, Global Knowledge Partnership on Migration and Development (KNOMAD), Washington, DC.

World Bank 2017, *Global Economic Prospects, June 2017: A Fragile Recovery*, Washington, DC, viewed 20 June 2017, https://openknowledge.worldbank.org/handle/10986/26800

10 Community forestry reinventing itself in Nepal

Robert Fisher, Richard Thwaites and Mohan Poudel

Introduction

The establishment of almost 20,000 local institutions in the form of CFUGS across rural Nepal is an extraordinary achievement. It is particularly impressive considering every one of these local institutions is uniquely established, based on negotiations around resource use and management between community members and the Department of Forests, that community forestry has improved forest quality and cover, and has increased the availability and access to forest products necessary for subsistence livelihoods. On a global scale these results are highly significant. Community forestry in Nepal is a genuine national program, far beyond a modest experiment in innovation.

Community forestry has also attempted to address poverty reduction and underlying social issues such as the empowerment of women and disadvantaged groups. As the chapters in this book have demonstrated, the outcomes of these efforts have been, at best, mixed. A major challenge has been the inequalities of power inherent in controlling valuable resources, both in terms of the power of central agencies compared to the power of decentralized communities and in terms of the unequal distribution of power within communities.

The progress of community forestry has continually faced challenges, many domestic and others arising from international and global change and influences. Since its early stages in the late 1970s it has continually evolved in response to changed priorities and in changing circumstances. This process of evolution has been an underlying theme of this book.

This concluding chapter will not present a chapter-by-chapter summary of the highlights on each of the themes covered by these chapters. Rather, the chapter will draw out some of the key themes and challenges arising from the chapters. It will look at the current context in which community forestry operates and attempt to draw out some of the new challenges for the future of community forestry.

The major themes that have emerged from the chapters in this book are the following:

- The relatively limited impacts of community forestry in addressing poverty and the needs of the marginalized;

- Issues of power evident both in the efforts of forest authorities to resist genuine devolved decision-making to communities and in the domination of CFUGs by local elites;
- The tendency of community forestry policies to be heavily influenced by international donors and the resulting tendency to add new tasks and responsibilities, such as those due to climate change adaptation and REDD+, onto Community Forest User Groups (CFUGs) in response to global agenda;
- The profound impact of economic and political change in Nepal and particularly the 'remittance economy';
- The need to establish more equitable and inclusive decision-making platforms for community forestry;
- The issue of the longer term relevance of and need for community forestry in the face of the changed global and national contexts in which it operates.

The achievements of community forestry: benefits to communities

The original aims of community forestry focused on improving the extent of forest cover as a response to extensive deforestation. The focus was on planting new forests. The earliest formal versions of community forestry were Panchayat Forests (related to plantation) and Panchayat Protected Forests (related to natural forests). Both of these programs were forest restoration and protection oriented. A concern with providing livelihood benefits to rural people, mainly in terms of meeting subsistence needs, became a focus in the late 1980s.

This section focuses on the benefits to communities. The environmental benefits have been covered in detail in Chapter 3 and are fairly clear. In general it is clear that community forestry has contributed to a considerable improvement in forest cover and condition in the areas in which it functions. There is also some (less clear) evidence of benefits to floral biodiversity.

The primary benefits of community forestry to communities have been in terms of contributions to rural livelihoods, primarily in terms of subsistence benefits (see Chapter 4). While there have been positive benefits, generalizations about these benefits need to be qualified. As is shown in Chapter 4, there is evidence that the poor have not always, or perhaps even not often, benefitted equally from availability of subsistence products (of which fuelwood is particularly important). Imposition of quotas may mean that poorer people have reduced access, because quotas are set by elite leaders who rely less on forests for tree products, which they may obtain from private land. In other words, some people may be absolutely worse off under user group rules.

As the ambitions of community forestry expanded, contributing to poverty reduction became a new goal. The addition of climate change adaptation and REDD+ as responsibilities for community forestry both included aspirations to contribute to poverty reduction. While the ambitions for poverty reduction have been well motivated, the achievements have not lived up to these ambitions. This is clear from the research described and the literature reviewed in Chapter 4.

Income generation has not reached the 'poorest of the poor', and despite require-ments for poverty reduction activities to be covered by user group budgets, the amounts provided are inadequate and are not available to all poor community members. Development funds from user group budgets are often spent on infra-structure such as roads and schools, which are of primary benefit to more wealthy community members.

In the case of REDD+ activities (Chapter 8), the amounts available for poverty reduction are also inadequate and, in any case, do not reach all people affected by REDD+ initiated restrictions on forest livelihoods and therefore do not address the need to compensate affected people for losses to livelihoods. There has been a failure to recognize the fact that serious compensation for the costs of imple-menting REDD+ and similar programs is a question of human rights. As new responsibilities such as those related to climate change, REDD+ and biodiversity are added to community forestry, the question arises of ensuring that fair and adequate compensation is provided to those who bear the costs of what are essen-tially services provided to the global community. Along with the human rights issue, this comes down to a question of legal tenure (Chapter 6). As long as rights are not secure, or are ignored, and their control over forest management decision-making is diluted, local communities will lose by trading away the right to harvest forest products on the expectation of their right to compensation from realizing value of other services derived from forest management, whether that value be carbon, biodiversity, water or any other service that might be valued beyond the confines of the local community.

It is clear from Chapters 4, 7 and 8 that the various activities of community forestry have had limited positive impact on the poor and have, in some cases, actually had negative impacts. As Chapter 4 suggests, there is a need for more direct targeting of the poor in order to achieve better outcomes. We suggest that questions remain as to what governance processes would achieve such targeting and what 'honest brokers' would facilitate it.

One further aspiration for community forestry has been the empowerment of women and various disadvantaged groups (such as the poor in general and disad-vantaged castes such as Dalits). It is important to note here that while women are generally disempowered, it must be recognized that they do not comprise a single group and that the category 'women' should be disaggregated to recognize the diverse situations of various women, such as wealthy women versus poor women, or high caste women versus lower caste women. Chapter 5 points out that user group decision-making processes do not provide for adequate inclusion of women or disadvantaged groups. This is an important point to emphasize but the obser-vation has been made widely in the literature and the need to empower these groups remains a major challenge. Empowerment of various groups of people is not an issue unique to community forestry in Nepal. It is shared with community-based natural resource management in general globally.

Although CFUGs have not been particularly successful in addressing equity issues, as discussed in Chapter 5, they have contributed to improved governance of natural resources and local development. Over the last two decades, in the absence of elected local government, they have contributed, to some extent, to filling a local vacuum of governance.

It is also important to note here that the findings of the various chapters in this book that deal with livelihoods, poverty and empowerment and the limited achievements of community forestry in addressing them are not new findings. They have been reported in a wide variety of literature (discussed in the chapters). We believe that the value of the chapters lies in the specific case studies that illustrate the processes underlying what is reported in the literature.

Power and decision-making at different scales

A clear threat to community forestry is the reluctance of actors at a variety of levels to relinquish control of forest resources to other actors. This pattern is very obvious in the reluctance of national forest authorities (and district-level Department of Forests staff) to support decentralized forest management in the form of devolved decision-making by user groups (see Chapters 2 and 9). It is also evident in the domination of user group decision-making by elites who frequently dominate the user group Executive Committees (see Chapters 4, 5, 7).

As is described in Chapter 2, the decentralization of forest management to CFUGs that was evident in the late 1980s, politically enabled by the revolution of 1990 and legislatively enabled by the 1993 forest legislation and subsequent regulations, was followed by efforts to impose new requirements for oversight by the Department of Forests. Under the Forest Act 1993, communities had a right to apply for community forest status, but forest management activities were subject to approval of an operational management plan by the District Forest Office. Later, requirements for preparation and approval of management plans became more onerous and required technical support and full surveys by the departmental staff, something which was quite time consuming.

These management plans were generally quite conservative in terms of forest use (Chapter 9), although in theory active management was permitted under the law and regulations. There were probably two reasons for this. The first is that the District Forest Office was reluctant to approve plans that allowed generous use of forest products. The second is that the local elites who dominated user group decision-making did not depend heavily on forest products and were willing to support conservative use. The reluctance of the Department of Forests to support community-operated saw mills (Chapter 2) is indicative of the extent to which the department has wished to control forest use.

The introduction of climate change and associated REDD+ activities, and of biodiversity into the community forestry agenda, has further contributed to the upward accountability of communities to government agencies. Meeting goals under these programs implied greater restrictions on forest utilization by communities, contributing to the limitations on the scope of community decision-making.

The new constitution of Nepal presents some challenges and opportunities for community forestry, although details of the way the new constitution will be implemented are not clear (Chapters 2 and 9). One of the key features of the constitution is that it aims to devolve much greater decision-making and budgetary control to the new local government institutions. This follows a period of nearly two decades without elected local government and when, in any case, the Village Development Committees had little budgetary role. The theory behind the constitution is that districts will have little or no political or administrative power, with government services being provided by officials located in the districts.

One obvious threat to community forestry is that the new local government institutions might impose further control over them or make political decisions that override the desires of CFUG members. According to an experienced commentator on community forestry in Nepal (pers. comm. to Fisher, April 2017), the alternative possibility would be that the local government could provide more direct financial and other support that is currently the responsibility of the Department of Forests. This informant also explained that the Ministry of Forests and Soil Conservation was arguing that forests and national parks should remain outside the decentralized structure and subject to the Ministry, which would mean that the forest administration would retain a high degree of control. While the outcome remains uncertain, this is a further clear example of reluctance to relinquish control.

The issue of contested control of forests and other natural resources, and the associated reality that control is contested at multiple scales, is not unique to community forestry in Nepal, but is widely prevalent. Ribot (2002) discusses the concept of 'democratic decentralization', arguing that much decentralization is not democratic decentralization. He shows how local institutions may often remain accountable to higher authorities, rather than downwardly accountable to the people they supposedly represent. He argues for the importance of downward accountability, from the leaders of local institutions towards their local constituents. This is remarkably consistent with the pattern evident in community forestry in Nepal, and the concept of 'downward accountability' is an obvious process to aim for. The challenge is what political process would achieve it and what would motivate those in control to hand over some of their power.

Donor influence and international agenda

International donors have influenced the direction of community forestry since its earliest days (see Chapters 2 and 9). The early emphasis on plantation was a response to the internationally prominent crisis narratives of the global fuelwood crisis and the Himalayan degradation crisis (see Chapter 2). One aspect of donor influence was the fact that donor funding is and has been a dominant component of forest sector funding, thus enabling it to influence priorities. Later, various elements of community forestry policy were promoted and supported by a variety of international multilateral and bilateral donors. Even the influential Master Plan for the Forestry Sector was implemented by an internationally funded consultant and the promotion of community forestry in the Master Plan was influenced by donor-funded projects. The introduction of the user group federation FECOFUN was supported by donor funding.

We believe, however, that it is important not to over emphasize the role of donor funding. Community forestry in Nepal was heavily influenced by local champions both in the forest administration and in the civil society that emerged particularly after the 1990 revolution. As pointed out in Chapter 2, there were many foresters (that is, inside the administration) who became deeply committed to community forestry as a movement. To ignore their influence would be to ignore the reality of individual agency inside Nepal in the context of community forestry. As pointed out in Chapter 2, the recognition and endorsement of the role of user groups was an outcome of the Decentralization Act 1987, and was therefore not a simple response to donor agendas.

International influence is evident in the emergence of new roles for CFUGs. Clear and relatively recent examples of this process are evident in the incorporation of climate change and REDD+ activities into community forestry activities and expectations of incorporation of greater responsibilities for biodiversity conservation. This influence has arisen from the global agenda and responses within Nepal have been supported by donor funding. As Chapters 7 and 8 show, the effects of such additions have complicated community forestry and, at least to some extent, worked against the interests of the poor and worked against decentralized decision-making, contributing to 'upward accountability'.

Clearly climate change is an important global issue and changes to climate reveal serious present and future threats to the lives of people in the rural areas. As CFUGs are widely distributed and have shown to be remarkably effective local institutions, especially during the political chaos of nearly two decades, community forestry obviously has an important role in adaptation to climate change.

Regardless of the appropriateness of CFUGs being called upon to take on new roles, an effect of this is that CFUGs are increasingly being expected to deal with too many goals, including sometimes competing goals. This is in itself a problem. It is complicated by a further issue, which is whether the current structures and processes of CFUG management are adequate to cover the increased complexity and whether more broadly based and inclusive institutions are required. The idea that new forms of institutions are needed to address the challenges arising from massive global and national changes is discussed in Chapter 9.

Political and economic change in Nepal

We have already discussed some of the political changes affecting Nepali society. It is clear that fundamental and very widespread changes are occurring. As Chapter 9 shows, the key linchpin underlying these changes is the remittance economy, which has resulted from several million Nepalis (mostly men) travelling overseas for employment. Significant amounts of the resulting income have been sent back to Nepal to support families. As Chapter 9 shows, the impacts of these remittances have included reduced dependence on agriculture, the increased role of women in remaining agriculture and reduced dependency on community forests for livelihood support. Other changes, not discussed in detail in Chapter 9, are the wide availability of communications (especially mobile phones) and a significant shift of elements of the population away from villages to regional towns.

Perhaps unsurprisingly, the benefits of the increased wealth have not been universal. As Chapter 9 shows, many people remain poor and disadvantaged. They would benefit from more use of forests in ways that support economic activities and income generation. Yet, paradoxically, despite community forestry having made some contributions to community governance and participation (as suggested in Chapter 8), CFUG committees (and thus, decision-making) remain dominated by elites who often have no active interest in forest management, and opportunities for poverty reduction are not often implemented.

This can be seen as a missed opportunity because there is a strong narrative within the forestry profession that sees that forests have been under-utilized in community forestry and argues for increased and sustainable use.

Conclusions: the future of community forestry and the global relevance of Nepal's experiences

It is clear from the chapters in this book that community forestry is becoming less relevant to some of its original purposes. Contributing to subsistence needs is much less relevant, at least to many households, than it was in the past. However, meeting subsistence needs is not the same as addressing poverty, which remains a major issue. A substantial contribution to forest restoration and conservation has been achieved and a system for managing forests for these purposes has been established. It seems almost certain that community forestry will need to adapt further to address new issues including climate change, but also including the need to make a greater contribution to economic development and empowerment of disadvantaged people.

The challenge is how to achieve that. The existing social structure and political system does not seem to be able to cope with the needs. Clearly, as far as community forestry is concerned, some sort of new institutional arrangements will need to develop that will be much more inclusive and incorporate a wider variety of actors in addition to community members, including academics, political leaders and activists (discussed in Chapter 9). The crucial role of these new institutions would be to think of community forestry as part of a wider economic, social and political system and attempt to understand that community forests operate in the context of the broad political economy.

The great challenge is where the impetus to create such new institutions will come from. Will entrenched interests continue to dominate? Who will be the 'honest brokers' who can facilitate the changes?

Does community forestry have a future? The chapters in this book have clearly shown that community forestry has been in a continual process of change and that it needs to adapt to new conditions if it is to remain relevant. Community forestry developed as a response to widely prevailing conditions in Nepal at a particular period. Many of those conditions have changed and will, no doubt, continue to change. Nevertheless, some level of dependence on forests for subsistence and poverty reduction is likely to continue for some time.

Community forestry is becoming less relevant to its original purposes, but still does have the potential to contribute to poverty reduction and development. With wider economic development and increasing urbanization, community forestry in anything closely resembling its current form may cease to be relevant. After all, to quote an aphorism originally attributed to Socrates, 'no (human) condition is permanent'. We strongly believe that, suitably adapted, it will remain relevant for the foreseeable future.

The global relevance of community forestry in Nepal

The experience of the community forestry movement in Nepal has wider importance. The extent to which it is practiced within Nepal is almost unique. Few community-based forestry programs have such wide application in their respective countries and few have achieved such wide development of local institutions. Much has been written about the program and some of the features have been applied in other contexts.

But the point of this book is that there are challenges. We believe these challenges need to be recognized and that they apply more widely than in Nepal.

The first challenge is to develop institutions that can continue to function without direct financial support and without continuing supervision. The CFUGs in Nepal largely achieved that during the civil war period. We suggest that part of the reason this occurred was because the CFUGs were supported by civil society and were not narrowly dependent on the forest administration. To a considerable extent CFUGs developed as part of a social movement.

The second challenge is to cultivate forest management institutions that truly represent the interests of the poor in which the decision-makers are 'downwardly accountable'.

The third challenge is to recognize that the role of forests in rural livelihoods is not static. This does not mean that community-based forest management is, or is rapidly becoming, irrelevant, but rather that it must adapt to changing conditions.

Reference

Ribot, J.C. 2002, *Democratic Decentralization of Natural Resources: Institutionalizing Popular Participation*, World Resources Institute, Washington, DC.

Index